Advances in MATHEMATICAL ECONOMICS

Aims and Scope. The project is to publish *Advances in Mathematical Economics* once a year under the auspices of the Research Center for Mathematical Economics. It is designed to bring together those mathematicians who are seriously interested in obtaining new challenging stimuli from economic theories and those economists who are seeking effective mathematical tools for their research.

The scope of *Advances in Mathematical Economics* includes, but is not limited to, the following fields:

- Economic theories in various fields based on rigorous mathematical reasoning.
- Mathematical methods (e.g., analysis, algebra, geometry, probability) motivated by economic theories.
- Mathematical results of potential relevance to economic theory.
- Historical study of mathematical economics.

Authors are asked to develop their original results as fully as possible and also to give a clear-cut expository overview of the problem under discussion. Consequently, we will also invite articles which might be considered too long for publication in journals.

More information about this series at http://www.springer.com/series/4129

Shigeo Kusuoka · Toru Maruyama
Editors

Advances in Mathematical Economics

Volume 21

Editors
Shigeo Kusuoka
Professor Emeritus
The University of Tokyo
Tokyo, Japan

Toru Maruyama
Professor Emeritus
Keio University
Tokyo, Japan

ISSN 1866-2226 ISSN 1866-2234 (electronic)
Advances in Mathematical Economics
ISBN 978-981-13-5061-0 ISBN 978-981-10-4145-7 (eBook)
DOI 10.1007/978-981-10-4145-7

Printed on acid-free paper

This Springer imprint is published by Springer Nature
The registered company is Springer Nature Singapore Pte Ltd.
The registered company address is: 152 Beach Road, #21-01/04 Gateway East, Singapore 189721, Singapore

Contents

Some Problems in Second Order Evolution Inclusions with Boundary Condition: A Variational Approach 1
Charles Castaing, Truong Le Xuan, Paul Raynaud de Fitte and Anna Salvadori

On Sufficiently-Diffused Information in Bayesian Games: A Dialectical Formalization 47
M. Ali Khan and Yongchao Zhang

On Supermartingale Problems 75
Shigeo Kusuoka

Bolza Optimal Control Problems with Linear Equations and Periodic Convex Integrands on Large Intervals 99
Alexander J. Zaslavski

Index 161

Some Problems in Second Order Evolution Inclusions with Boundary Condition: A Variational Approach

Charles Castaing, Truong Le Xuan, Paul Raynaud de Fitte and Anna Salvadori

Abstract We prove, under appropriate assumptions, the existence of solutions for a second order evolution inclusion with boundary conditions via a variational approach.

Keywords Bounded variation · Epiconvergence · Biting Lemma · Subdifferential · Young measures

Article type: Research Article
Received: September 12, 2016
Revised: January 6, 2017

JEL Classification: C61, C73.
Mathematics Subject Classifications (2010): 34A60, 34B15.

C. Castaing (✉)
Département de Mathématiques, Université Montpellier II, Case 051, 34095 Montpellier Cedex 5, France
e-mail: castaing.charles@gmail.com

T. Le Xuan
Department of Mathematics and Statistics, University of Economics of Ho Chi Minh City, 59C Nguyen Dinh Chieu Str. Dist. 3, Ho Chi Minh City, Vietnam
e-mail: lxuantruong@gmail.com

P. Raynaud de Fitte
Laboratoire Raphaël Salem, UMR CNRS 6085, Normandie Université, Rouen, France
e-mail: prf@univ-rouen.fr

A. Salvadori
Dipartimento di Matematica, Università'di Perugia, via Vanvitelli 1, 06123 Perugia, Italy
e-mail: mateas@unipg.it

S. Kusuoka and T. Maruyama (eds.), *Advances in Mathematical Economics*, Advances in Mathematical Economics 21,
DOI 10.1007/978-981-10-4145-7_1

1 Introduction

In the present paper, we prove, under appropriate assumptions, the existence of solutions for a second order evolution inclusion with boundary conditions governed by subdifferential operators of the form

$$f(t) \in \ddot{u}(t) + M\dot{u}(t) + \partial\varphi(u(t)), t \in [0, T]. \tag{I}$$

Here, M is positive, φ is a lower semicontinuous convex proper function defined on $\mathbf{R}^d$ and $\partial\varphi(u(t))$ is the subdifferential of the function φ at the point $u(t)$ and the perturbation f belongs to $L^2_{\mathbf{R}^d}([0, T])$. It is well known that this problem is difficult and needs a specific treatment via the Moreau-Yosida approximation or epiconvergence approach. See Attouch–Cabot–Redon [4] and Schatzmann [24] for a deep study of these problems, Castaing–Raynaud de Fitte–Salvadori [11], Castaing–Le Xuan Truong [8] dealing with second order evolution with m-point boundary conditions via the epiconvergence approach. These considerations lead us to consider the variational limits of a fairly general approximating problem

$$f^n(t) \in \ddot{u}^n(t) + M\dot{u}^n(t) + \partial\varphi_n(u^n(t)), t \in [0, T] \tag{II}$$

where u^n is a $W^{2,1}_{\mathbf{R}^d}([0, T])$-solution, f^n weakly converging in $L^2_{\mathbf{R}^d}([0, T])$ to f^∞, φ_n is a convex Lipschitz function which epiconverges to a lower semicontinuous convex proper function φ_∞. This approximating problem covers various type of problems of practical interest in several dynamic systems, evolution inclusion, control theory etc. Here we focus on several variational limits of solutions via the Biting Lemma and Young measures and other tools occurring in this approach by showing under suitable limit assumption on the boundary conditions that $(\ddot{u}^n)$ is $L^1_{\mathbf{R}^d}([0, T])$-bounded. This main fact allows to study the variational limit of solutions in this problem, in particular, the traditional estimated energy for the variational limit solutions is conserved almost everywhere. The applicability of our abstract framework given therein (Proposition 3.3) will be exemplified in considering the existence of solution for second order differential inclusions

$$f(t) \in \ddot{u}(t) + M\dot{u}(t) + \partial\varphi(u(t)), t \in [0, T]$$

under m-point boundary condition or anti-periodic conditions and further related second order evolution inclusions in the literature. This will be done by applying our abstract result to the single valued approximating problem

$$f^n(t) = \ddot{u}^n(t) + M\dot{u}^n(t) + \nabla\varphi_n(u^n(t)), t \in [0, T] \tag{III}$$

where $\nabla\varphi_n$ is the gradient of the C^1, Lipschitz, convex function φ_n that epi-converges to a proper convex lower semicontinuous function φ_∞ and f^n weakly converges in $L^2_{\mathbf{R}^d}([0, T])$ to f^∞ so that the variational limit solutions u^∞ to (III) are *generalized solutions* to the inclusion

$$f^\infty(t) \in \ddot{u}^\infty(t) + M\dot{u}^\infty(t) + \partial\varphi_\infty(u^\infty(t)), t \in [0, T]$$

with appropriate properties, namely, the solution limit u^∞ is $W_{BV}^{1,1}([0, T])$, that is, u^∞ is continuous and its derivative $\dot{u}^\infty$ is bounded variation (BV for short) and the estimated energy holds almost everywhere

$$\varphi_\infty(u^\infty(t)) + \frac{1}{2}||\dot{u}^\infty(t)||^2 = \varphi_\infty(u_0) + \frac{1}{2}||\dot{u}_0)||^2 - M\int_0^t ||\dot{u}^\infty(s)||^2 ds + \int_0^t \langle f^\infty(s), \dot{u}^\infty(s)\rangle ds$$

with further related variational inclusion, in particular,

$$f^\infty(t) \in \zeta^\infty(t) + Mu^\infty(t) + \partial\varphi_\infty(u^\infty(t)), t \in [0, T]$$

almost everywhere, ζ^∞ being the biting limit of the $L^1_{\mathbf{R}^d}([0, T])$-bounded sequence $(\ddot{u}^n)$. Section 3 is devoted to second order evolution inclusion with boundary conditions. We present the variational limits of the general approximating problem (II) and the applications of variational limits of the approximating problem (III) to the existence problem of second order evolution inclusion (I) involving variational techniques, the Biting Lemma, the characterization of the second dual of $L^1_{\mathbf{R}^d}$ and Young measures. It is worth to mention that the approximation (III) occurs in practical cases of second order evolution inclusion governed by subdifferential operators. For instance, Attouch–Cabot–Redon [4] considered the approximating problem

$$0 = \ddot{u}^n(t) + \gamma\dot{u}^n(t) + \nabla\varphi_n(u^n(t)), t \in [0, T]$$

$$u^n(0) = u_0^n, \dot{u}^n(0) = \dot{u}_1^n$$

where γ is positive, $\nabla\varphi_n$ is the gradient of a C^1, smooth function. Schatzmann [24] considered the approximating problem

$$f(t) = \ddot{u}_\lambda(t) + \partial\varphi_\lambda(u_\lambda(t)), t \in [0, T]$$

$$u_\lambda(0) = u_0, \dot{u}_\lambda(0) = u_1$$

where $f \in L^2_{\mathbf{R}^d}([0, T])$ and $\partial\varphi_\lambda$ is the Moreau-Yosida approximation to the lower semicontinuous convex proper function φ. M. Mabrouk [19] continued the work of M. Schatzmann [24] by considering the approximating problem

$$f_\lambda(t) = \ddot{u}_\lambda(t) + \nabla\varphi_\lambda(u_\lambda(t)), t \in [0, T]$$
$$u_\lambda(0) = u_0, \dot{u}_\lambda(0) = u_1,$$

with $f_\lambda \in L^1_{\mathbf{R}^d}([0, T])$. In Sect. 4, we apply our techniques to the study of both first order and second order evolution equations with anti-periodic boundary condition using the approximating problem

$$\begin{gathered} f^n(t) = \ddot{u}^n(t) + M\dot{u}^n(t) + \nabla\varphi_n(u^n(t)), t \in [0, T] \\ u^n(0) = -u^n(T), \end{gathered}$$

where $u^n \in W^{2,2}_{\mathbf{R}^d}([0, T])$ and $f^n \in L^2_{\mathbf{R}^d}([0, T])$, see H. Okochi [22], A. Haraux [17], Aftabizadeh, Aizicovici and Pavel [1, 2], Aizicovici and Pavel [3] and the references therein.

A general analysis of some related problems in Hilbert space is available, c.f K. Maruo [19] and M. Schatzmann [24].

2 Some Existence Theorems in Second Order Evolution Inclusions with *m*-Point Boundary Condition

We will use the following definitions and notations and summarize some basic results.

- Let E be a separable Banach space, $\overline{B}_E(0, 1)$ is the closed unit ball of E.
- $c(E)$ (resp. $cc(E)$) (resp. $ck(E)$)(resp. $cwk(E)$) is the collection of nonempty closed (resp. closed convex) (resp. compact convex) (resp. weakly compact convex) subsets of E.
- If A is a subset of E, $\delta^*(., A)$ is the support function of A.
- $\mathcal{L}([0, T])$ is the σ-algebra of Lebesgue measurable subsets of $[0, T]$.
- If X is a topological space, $\mathcal{B}(X)$ is the Borel tribe of X.
- $L^1_E([0, T], dt)$ (shortly $L^1_E([0, T])$) is the Banach space of Lebesgue–Bochner integrable functions $f : [0, T] \to E$.
- A mapping $u : [0, T] \to E$ is *absolutely continuous* if there is a function $\dot{u} \in L^1_E([0, T])$ such that $u(t) = u(0) + \int_0^t \dot{u}(s)\, ds, \ \forall t \in [0, T]$.
- If X is a topological space, $\mathcal{C}_E(X)$ is the space of continuous mappings $u : X \to E$ equipped with the norm of uniform convergence.
- A set-valued mapping $F : [0, T] \rightrightarrows E$ is measurable if its graph belongs to $\mathcal{L}([0, T]) \otimes \mathcal{B}(E)$.
- A convex weakly compact valued mapping $F : X \to ck(E)$ defined on a topological space X is scalarly upper semicontinuous if for every $x^* \in E^*$, the scalar function $\delta^*(x^*, F(.))$ is upper semicontinuous on X.

We refer to [13] for measurable multifunctions and Convex Analysis.

For the sake of completeness, we recall and summarize some results developed in [9]. By $W^{2,1}_E([0, T])$ we denote the set of all continuous functions in $\mathcal{C}_E([0, T])$ such that their first derivatives are continuous and their second derivatives belong to $L^1_E([0, T])$.

Lemma 2.1 *Assume that E is a separable Banach space. Let $0 < \eta_1 < \eta_2 < \cdots < \eta_{m-2} < 1$, $\gamma > 0$, $m > 3$ be an integer number, and $\alpha_i \in \mathbf{R}$ $(i = 1, \ldots, m-2)$ satisfying the condition*

$$\sum_{i=1}^{m-2} \alpha_i - 1 + \exp(-\gamma) - \sum_{i=1}^{m-2} \alpha_i \exp(-\gamma\eta_i)) \neq 0.$$

Let $G : [0, 1] \times [0, 1] \to \mathbf{R}$ be the function defined by

$$G(t, s) = \begin{cases} \frac{1}{\gamma}(1 - \exp(-\gamma(t - s))), & 0 \leq s \leq t \leq 1 \\ 0, & t < s \leq 1 \end{cases} + \frac{A}{\gamma}(1 - \exp(-\gamma t))\,\phi(s), \tag{2.1}$$

where

$$\phi(s) = \begin{cases} 1 - \exp(-\gamma(1 - s)) - \sum_{i=1}^{m-2} \alpha_i (1 - \exp(-\gamma(\eta_i - s))), & 0 \leq s < \eta_1, \\ 1 - \exp(-\gamma(1 - s)) - \sum_{i=2}^{m-2} \alpha_i (1 - \exp(-\gamma(\eta_i - s))), & \eta_1 \leq s \leq \eta_2, \\ \ldots\ldots & \\ 1 - \exp(-\gamma(1 - s)), & \eta_{m-2} \leq s \leq 1, \end{cases} \tag{2.2}$$

and

$$A = \left(\sum_{i=1}^{m-2} \alpha_i - 1 + \exp(-\gamma) - \sum_{i=1}^{m-2} \alpha_i \exp(-\gamma\eta_i)\right)^{-1}. \tag{2.3}$$

Then the following assertions hold

(i) For every fixed $s \in [0, 1]$, the function $G(., s)$ is right derivable on $[0, 1[$ and left derivable on $]0, 1]$. Its derivative is given by

$$\left(\frac{\partial G}{\partial t}\right)_+ (t, s) = \begin{cases} \exp(-\gamma(t - s)), & 0 \leq s \leq t < 1 \\ 0, & 0 \leq t < s < 1 \end{cases} + A\exp(-\gamma t)\phi(s), \tag{2.4}$$

$$\left(\frac{\partial G_\tau}{\partial t}\right)_- (t, s) = \begin{cases} \exp(-\gamma(t - s)), & 0 \leq s < t \leq 1 \\ 0, & 0 < t \leq s \leq 1 \end{cases} + A\exp(-\gamma t)\phi(s). \tag{2.5}$$

(ii) $G(\cdot, \cdot)$ and $\frac{\partial G}{\partial t}(\cdot, \cdot)$ satisfies

$$|G(t, s)| \leq M_G \quad \text{and} \quad \left|\frac{\partial G}{\partial t}(t, s)\right| \leq M_G \quad \forall (t, s) \in [0, 1] \times [0, 1],$$

where

$$M_G = \max\{\gamma^{-1}, 1\}\left[1 + |A|\left(1 + \sum_{i=1}^{m-2} |\alpha_i|\right)\right].$$

(iii) *If* $u \in W_E^{2,1}([0,1])$ *with* $u(0) = x$ *and* $u(1) = \sum_{i=1}^{m-2} \alpha_i u(\eta_i)$, *then*

$$u(t) = e_x(t) + \int_0^1 G(t,s)(\ddot{u}(s) + \gamma \dot{u}(s))ds, \quad \forall t \in [0,1],$$

where

$$e_x(t) = x + A\left(1 - \sum_{i=1}^{m-2} \alpha_i\right)(1 - \exp(-\gamma t))x.$$

(iv) *Let* $f \in L_E^1([0,1])$ *and let* $u_f : [0,1] \to E$ *be the function defined by*

$$u_f(t) = e_x(t) + \int_0^1 G(t,s) f(s)ds \quad \forall t \in [0,1].$$

Then we have

$$u_f(0) = x \quad u_f(1) = \sum_{i=1}^{m-2} \alpha_i u_f(\eta_i).$$

Further the function u_f *is weakly derivable on* $[0,1]$ *and its weak derivative* $\dot{u}_f$ *is defined by*

$$\dot{u}_f(t) = \lim_{h \to 0} \frac{u_f(t+h) - u_f(t)}{h} = \dot{e}_x(t) + \int_T^1 \frac{\partial G}{\partial t}(t,s) f(s)ds,$$

with

$$\dot{e}_x(t) = \gamma A\left(1 - \sum_{i=1}^{m-2} \alpha_i\right)\exp(-\gamma t)x.$$

(v) *If* $f \in L_E^1([0,1])$, *the function* $\dot{u}_f$ *is weakly derivable, and its weak derivative* $\ddot{u}_f$ *satisfies*

$$\ddot{u}_f(t) + \gamma \dot{u}_f(t) = f(t) \quad \text{a.e. } t \in [0,1].$$

The following is a direct consequence of Lemma 2.1.

Proposition 2.1 *Let* $f \in L_E^1([0,1])$. *The m-point boundary problem*

$$\begin{cases} \ddot{u}(t) + \gamma \dot{u}(t) = f(t), \ t \in [0,1] \\ u(0) = x, u(1) = \sum_{i=1}^{m-2} \alpha_i u(\eta_i) \end{cases}$$

has a unique $W_E^{2,1}([0,1])$*-solution* u_f, *with integral representation formulas*

$$\begin{cases} u_f(t) = e_x(t) + \int_0^1 G(t,s) f(s)ds, \ t \in [0,1] \\ \dot{u}_f(t) = \dot{e}_x(t) + \int_0^1 \frac{\partial G}{\partial t}(t,s) f(s)ds, \ t \in [0,1]. \end{cases}$$

where

$$\begin{cases} e_x(t) = x + A(1 - \sum_{i=1}^{m-2} \alpha_i)(1 - \exp(-\gamma t))x \\ \dot{e}_x(t) = \gamma A \left(1 - \sum_{i=1}^{m-2} \alpha_i\right) \exp(-\gamma t)x \\ A \quad = \left(\sum_{i=1}^{m-2} \alpha_i - 1 + \exp(-\gamma) - \sum_{i=1}^{m-2} \alpha_i \exp(-\gamma(\eta_i))\right)^{-1}. \end{cases}$$

The following result and its notation will be used in the next section.

Proposition 2.2 *With the hypotheses and notations of Proposition 2.1, let E be a separable Banach space and let $X : [0, 1] \rightrightarrows E$ be a measurable convex weakly compact valued and integrably bounded mapping. Then the solution set of $W_E^{2,1}([0, 1])$-solutions to*

$$\begin{cases} \ddot{u}_f(t) + \gamma \dot{u}_f(t) = f(t), \ f \in S_X^1 \\ u_f(0) = x, \quad u_f(1) = \sum_{i=1}^{m-2} \alpha_i u_f(\eta_i) \end{cases}$$

is bounded, convex, equicontinuous and sequentially weakly compact in $\mathcal{C}_E([0, 1])$.

Proof Let us set

$$\mathcal{X} := \left\{u_f \in \mathcal{C}_E([0, 1]) : u_f(t) = e_x(t) + \int_0^1 G(t, s) f(s) ds, \ t \in [0, 1], \ f \in S_X^1\right\}$$

with

$$\begin{cases} e_x(t) = x + A(1 - \sum_{i=1}^{m-2} \alpha_i)(1 - \exp(-\gamma t))x, \ t \in [0, 1] \\ \dot{e}_x(t) = \gamma A \left(1 - \sum_{i=1}^{m-2} \alpha_i\right) \exp(-\gamma t)x, \ t \in [0, 1] \\ A \quad = \left(\sum_{i=1}^{m-2} \alpha_i - 1 + \exp(-\gamma) - \sum_{i=1}^{m-2} \alpha_i \exp(-\gamma(\eta_i))\right)^{-1}. \end{cases}$$

Taking account of the properties of G in Lemma 2.1, it is not difficult to show that $\mathcal{X}$ is bounded, convex, equicontinuous and relatively weakly compact in $\mathcal{C}_E([0, 1])$ because for each $t \in [0, T]$, $\int_0^1 G(t, s)X(s)ds$ is convex and weakly compact, see e.g. [11]. We only need to check the compactness property since other properties are obvious. Indeed, let $u_{f_n} \in \mathcal{X}$. As S_X^1 is $\sigma(L_E^1, L_{E_s^*}^\infty)$ sequentially compact, we may assume that (f_n) $\sigma(L_E^1, L_{E_s^*}^\infty)$ converges to $f_\infty \in S_X^1$. Then we have for each $t \in [0, 1]$,

$$\begin{aligned} \mathrm{w-}\lim_n u_{f_n}(t) &= e_x(t) + \mathrm{w} - \lim_n \int_0^1 G(t, s) f_n(s) ds \\ &= e_x(t) + \int_0^1 G(t, s) f_\infty(s) ds := u_{f_\infty}(t). \end{aligned}$$

This means that $u_{f_n}(t)$ converges to $u_{f_\infty}(t)$ in E_σ for every $t \in [0, 1]$. Hence u_{f_n} converges weakly in $\mathcal{C}_E([0, 1])$ to $u_{f_\infty} \in \mathcal{X}$. Similarly using the properties of $\frac{\partial G}{\partial t}$ in Lemma 2.1,

$$\mathcal{Y} := \left\{ \dot{u}_f \in \mathcal{C}_E([0, 1] : \dot{u}_f(t) = \dot{e}_x(t) + \int_0^1 \frac{\partial G}{\partial t}(t, s) f(s) ds, \ t \in [0, 1], \ f \in S^1_X \right\}$$

is bounded, convex, equicontinuous and sequentially weakly compact in $\mathcal{C}_E([0, 1])$ with

$$\begin{cases} \dot{e}_x(t) = \gamma A \left(1 - \sum_{i=1}^{m-2} \alpha_i\right) \exp(-\gamma t) x, \ t \in [0, 1] \\ A = \left(\sum_{i=1}^{m-2} \alpha_i - 1 + \exp(-\gamma) - \sum_{i=1}^{m-2} \alpha_i \exp(-\gamma(\eta_i))\right)^{-1}, \end{cases}$$

and we have

$$\begin{aligned} \text{w} - \lim_n \dot{u}_{f_n}(t) &= \dot{e}_x(t) + \text{w} - \lim_n \int_0^1 \frac{\partial G}{\partial t}(t, s) f_n(s) ds \\ &= \dot{e}_x(t) + \int_0^1 \frac{\partial G}{\partial t}(t, s) f_\infty(s) ds := u_{f_\infty}(t). \end{aligned}$$

This means that $\dot{u}_{f_n}(t)$ converges to $\dot{u}_{f_\infty}(t)$ in E_σ for every $t \in [0, 1]$. ■

Remark In the context of Control Theory, we have stated in the proof of Proposition 2.2, the dependence of the solution with respect to the control $f \in S^1_X$. Namely, if u_{f_n} is the $W^{2,1}_E([0, 1])$-solution to

$$\begin{cases} \ddot{u}_{f_n}(t) + \gamma \dot{u}_{f_n}(t) = f_n(t), \quad t \in [0, 1] \\ u_{f_n}(0) = x, \quad u_{f_n}(1) = \sum_{i=1}^{m-2} \alpha_i u_{f_n}(\eta_i) \end{cases}$$

and if (f_n) converges $\sigma(L^1_E, L^\infty_{E^*_s})$ to $f_\infty \in S^1_X$, then $(u_{f_n}(t))$ converges to $u_{f_\infty}(t)$ and $(\dot{u}_{f_n}(t))$ converges to $\dot{u}_{f_\infty}(t)$, in E_σ for every $t \in [0, 1]$ where u_{f_∞} is the $W^{2,1}_E([0, 1])$-solution to

$$\begin{cases} \ddot{u}_{f_\infty}(t) + \gamma \dot{u}_{f_\infty}(t) = f_\infty(t), \quad t \in [0, 1] \\ u_{f_\infty}(0) = x, \quad u_{f_\infty}(1) = \sum_{i=1}^{m-2} \alpha_i u_{f_\infty}(\eta_i). \end{cases}$$

The above remark is of importance since it allows to prove further results. Here is an application to the existence of $W^{2,1}_E([0, 1])$-solution to a second order differential inclusion with m-point boundary condition.

Proposition 2.3 *Let $X : [0, 1] \rightrightarrows E$ be a convex weakly compact valued measurable and integrably bounded mapping, $F : [0, 1] \times E \times E \rightrightarrows E$ be a convex weakly compact valued mapping satisfying*

(1) *For each* $x^* \in E^*$, *the scalar function* $\delta^*(x^*, F(., ., .))$ *is* $\mathcal{L}_\lambda([0,1]) \otimes \mathcal{B}(E_\sigma) \otimes \mathcal{B}(E_\sigma)$*-measurable,*[1]
(2) *For each* $x^* \in E^*$ *and for each* $t \in [0,1]$, *the scalar function* $\delta^*(x^*, F(t, ., .))$ *is sequentially weakly upper semicontinuous, i.e., for any sequence* (x_n) *in* E *weakly converging to* $x \in E$, *for any sequence* (y_n) *in* E *weakly converging to* $y \in E$, $\limsup_n \delta^*(x^*, F(t, x_n, y_n)) \leq \delta^*(x^*, F(t, x, y))$,
(3) $F(t, x, y) \in X(t)$ *for all* $(t, x, y) \in [0,1] \times E \times E$.
Then the $W_E^{2,1}([0,1])$*-solutions set to*

$$\begin{cases} \ddot{u}(t) + \gamma \dot{u}(t) \in F(t, u(t), \dot{u}(t))), \; t \in [0,1] \\ u(0) = x, \quad u(1) = \sum_{i=1}^{m-2} \alpha_i u(\eta_i) \end{cases}$$

is non empty and weakly compact in the space $C_E([0,1])$.

Proof The sets

$$\mathcal{X} := \left\{ u_f \in \mathcal{C}_E([0,1]) : u_f(t) = e_x(t) + \int_0^1 G(t,s) f(s) ds, \; f \in S_X^1, \; t \in [0,1] \right\} \tag{2.3.1}$$

and

$$\mathcal{Y} := \left\{ \dot{u}_f \in \mathcal{C}_E([0,1]) : \dot{u}_f(t) = \dot{e}_x(t) + \int_0^1 \frac{\partial G}{\partial t}(t,s) f(s) ds, \; t \in [0,1], \; f \in S_X^1 \right\} \tag{2.3.2}$$

are bounded, convex, equicontinuous and weakly compact in $C_E([0,1])$. By condition (3), it is clear that

$$F(t, u_f(t), \dot{u}_f(t)) \subset X(t) \tag{2.3.4}$$

for all $t \in [0,1]$ and for all $f \in S_X^1$. Further, recall that S_X^1 is $\sigma(L_E^1, L_{E^*}^\infty)$-compact (see e.g. [10]). Using (1)–(3), for each $f \in S_X^1$, let us consider the convex $\sigma(L_E^1, L_{E^*}^\infty)$-compact valued mapping $\Psi : S_X^1 \rightrightarrows S_X^1$ defined by

$$\Psi(f) := \{g \in S_X^1 : g(t) \in F(t, u_f(t), \dot{u}_f(t)), \text{ a.e. } t \in [0,1]\}.$$

Now we are going to show that Ψ is upper semi continuous on the convex $\sigma(L_E^1, L_{E^*}^\infty)$-compact set S_X^1. We need to check that the graph of Ψ is $\sigma(L_E^1, L_{E^*}^\infty)$-closed in $S_X^1 \times S_X^1$. Let $g_n \in \Psi(f_n)$ such that f_n, $\sigma(L_E^1, L_{E^*}^\infty)$-converges to $f \in S_X^1$ and g_n $\sigma(L_E^1, L_{E^*}^\infty)$-converges to $g \in S_X^1$. By compactness of $\mathcal{X}$ and $\mathcal{Y}$, it follows that $u_{f_n}(t) \to u_f(t)$ in E_σ and $\dot{u}_{f_n}(t) \to \dot{u}_f(t)$ in E_σ for every $t \in [0,1]$. From the inclusion $g_n \in \Psi(f_n)$, we have, for each $x^* \in E^*$ and for each $A \in \mathcal{L}_\lambda([0,1])$

[1] Actually $\mathcal{B}(E_\sigma) = \mathcal{B}(E)$ since E is separable.

$$\langle 1_A(t)x^*, g_n(t)\rangle \leq 1_A(t)\delta^*(x^*, F(t, u_{f_n}(t), \dot{u}_{f_n}(t))),$$

so that, by integration,

$$\int_A \langle x^*, g_n(t)\rangle dt \leq \int_A \langle x^*, F(t, u_{f_n}(t), \dot{u}_{f_n}(t))\rangle dt.$$

We thus have

$$\begin{aligned}\int_A \langle x^*, g(t)\rangle dt &= \lim_n \int_A \langle x^*, g_n(t)\rangle dt \\ &\leq \limsup_n \int_A \delta^*(x^*, F(t, u_{f_n}(t), \dot{u}_{f_n}(t))dt \\ &\leq \int_A \delta^*(x^*, F(t, u_f(t), \dot{u}_f(t)))dt.\end{aligned}$$

Whence we get

$$\int_A \langle x^*, g(t)\rangle dt \leq \int_A \delta^*(x^*, F(t, u_f(t), \dot{u}_f(t))dt$$

for every $A \in \mathcal{L}_\lambda([0,1])$. Consequently

$$\langle x^*, g(t)\rangle \leq \delta^*(x^*, F(t, u_f(t), \dot{u}_f(t)) \text{ a.e.}$$

Taking a dense sequence (e_k^*) in E^* with respect to the Mackey topology $\tau(E^*, E)$, we get

$$\langle e_k^*, g(t)\rangle \leq \delta^*(e_k^*, F(t, u_f(t), \dot{u}_f(t)) \text{ a.e.}$$

for all $k \in \mathbf{N}$. By [13, Proposition III.35], we get finally

$$g(t) \in F(t, u_f(t), \dot{u}_f(t))) \text{ a.e.}$$

which proves that $g\ in \Psi(f)$. Whence by Kakutani-Ky Fan fixed point theorem Ψ admits a fixed point $f \in S^1_X$. This is a solution to the second order differential inclusion under consideration. Using Lemma 2.1, such a fixed point f verifies

$$\begin{cases} \ddot{u}_f(t) + \gamma \dot{u}_f(t) \in F(t, u_f(t), \dot{u}_f(t)), \text{ a.e. } t \in [0,1] \\ u_f(0) = x, \quad u_f(1) = \sum_{i=1}^{m-2} \alpha_i u(\eta_i). \end{cases}$$

The compactness of the solution set follows from the compactness of $\mathcal{X}$. ■

Second Order Evolution Inclusions Governed by Subdifferential Operators

We need to recall and summarize some notions on the subdifferential mapping of local Lipchitz functions developed by L. Thibault [25]. Let E be a separable Banach space. Let $f : E \to \mathbf{R}$ be a locally Lipschitz function. By Christensen [14, Theorem 7.5], there is a set D_f such that its complementary is Haar-nul (hence D_f is dense in E) such that for all $x \in D_f$ and for all $v \in E$

$$r_f(x, v) = \lim_{\delta \to 0} \frac{f(x + \delta v) - f(x)}{\delta}$$

exists and $v \mapsto r_f(x, v)$ is linear and continuous. Let us set $\nabla f(x) = r_f(x, .) \in E^*$. Then $r_f(x, v) = \langle \nabla f(x), v \rangle$, $\nabla f(x)$ is the gradient of f at the point x. Let us set

$$\mathcal{L}_f(x) = \{ \lim_{j \to \infty} \nabla f(x_j) | x_j \in D_f, x_j \to x\}.$$

By definition, the subdifferential $\partial f(x)$ in the sense of Clarke [15] at the point $x \in E$ is defined by

$$\partial f(x) = \overline{co}\, \mathcal{L}_f(x).$$

The generalized directional derivative of f at a point $x \in E$ in the direction $v \in E$ is denoted by

$$f^{\cdot}(x, v) = \limsup_{h \to 0, \delta \to 0} \frac{f(x + h + \delta v) - f(x + h)}{\delta}.$$

Proposition 2.4 *Let $f : E \to \mathbf{R}$ be a locally Lipchitz function. Then the subdifferential $\partial f(x)$ at the point $x \in E$ is convex weak star compact and*

$$f^{\cdot}(x, v) = \sup\{\langle \zeta^*, v \rangle | \zeta^* \in \partial f(x)\} \quad \forall v \in E$$

that is, $f^{\cdot}(x, .)$ is the support function of $\partial f(x)$.

Proof See Thibault [25, Proposition I.12]. ■

Here are some useful properties of the subdifferential mapping.

Proposition 2.5 *Let $f : E \to \mathbf{R}$ be a locally Lipchitz function. Then the convex weak star compact valued subdifferential mapping ∂f is upper semicontinuous with respect to the weak star topology.*

Proof See [25, Proposition I.17]. Indeed we have

$$\delta^*(v, \partial f(x)) = f^{\cdot}(x; v) = \limsup_{h \to 0, \delta \to 0} \frac{[f(x + h + \delta v) - f(x + h)]}{\delta}.$$

As $f^{\cdot}(.; v)$ is upper semicontinuous and ∂f is convex compact valued in E^*_s, by [13], ∂f is upper semicontinuous in E^*_s. ■

Proposition 2.6 *Let $(T, \mathcal{T})$ a measurable space, and let $f : T \times E \to \mathbf{R}$ such that $f(., \zeta)$ is $\mathcal{T}$-measurable, for every $\zeta \in E$.*

$f(t,.)$ is locally Lipschitz for every $t \in T$.
Let $f_t^{\cdot}(x; v)$ be the directional derivative of $f(t,.) := f_t$ in the direction v for every fixed $t \in T$. Let x and v be two $\mathcal{T}$-measurable mappings from T to E. Then the following hold:

(a) the mapping $t \mapsto f_t^{\cdot}(x(t); v(t))$ is $\mathcal{T}$-measurable.
(b) the mapping $t \mapsto \partial f_t(x(t))$ is graph measurable, that is, its graph belongs to $\mathcal{T} \otimes \mathcal{B}(E_s^*)$.

Proof See Thibault [25, Proposition I.20 and Corollary I.21]. Note that the convex weak star compact valued mapping $t \mapsto \partial f_t(x(t))$ is scalarly $\mathcal{T}$-measurable, and so enjoys good measurability properties because E_s^* is a locally convex Lusin space. ■

We begin with a second order differential inclusion involving the subdifferential operator.

Proposition 2.7 *Assume that $E = \mathbf{R}^d$, and $h : [0, 1] \times \mathbf{R}^d \times \mathbf{R}^d \to \mathbf{R}^d$ be a bounded Carathéodory mapping, that is, h is separately Lebesque-measurable on $[0, 1]$, separately continuous on $\mathbf{R}^d \times \mathbf{R}^d$, $||h(t, x, y)|| \leq \alpha(t)$, $\forall (t, x, y) \in [0, T] \times \mathbf{R}^d \times \mathbf{R}^d$ where α is positive Lebesque-integrable. Let $f : [0, 1] \times E \to \mathbf{R}$ be a mapping such that*

(1) $\forall x \in E$, $f(., x)$ is Lebesgue-measurable,
(2) There exists $\beta \in L^1_{\mathbf{R}^+}([0, 1])$ such that, for all $t \in [0, 1]$, for all $x, y \in E$,

$$||f(t, x) - f(t, y)|| \leq \beta(t)||x - y||.$$

Then the following hold

(a) $\partial f_t(x) \subset \beta(t)\overline{B}_E$, for all $(t, x) \in [0, 1] \times E$,
(b) The $W_E^{2,1}([0, 1])$-solution set to

$$\begin{cases} \ddot{u}(t) + \gamma\dot{u}(t) \in \partial f_t(u(t)) + h(t, u(t), \dot{u}(t)), \text{ a.e. } t \in [0, 1] \\ u(0) = x, \quad u(1) = \sum_{i=1}^{m-2} \alpha_i u(\eta_i) \end{cases}$$

is compact in the space $\mathcal{C}_E([0, T])$.

Proof The proof is immediate by applying Proposition 2.3 to the convex compact valued mapping $(t, x, y) \mapsto \partial f_t(x) + h(t, x, y)$, taking account of the properties of the subdifferential mapping and its measurable properties given in Proposition 2.6. ■

We finish this section with a variant which has some importance in the study of epiconvergence problem for the approximating system

$$\ddot{u}(t) + \gamma\dot{u}(t) = h(t, u(t), \dot{u}(t)) - \nabla\varphi(u(t))$$

where φ is C^1 and Lipschitz.

Proposition 2.8 *Assume that $E = \mathbf{R}^d$, $\varphi : E \to \mathbf{R}$ is C^1, Lipschitz, and that $h : [0,1] \times \mathbf{R}^d \times \mathbf{R}^d \to \mathbf{R}^d$ is a bounded Carathéodory mapping, that is, h is separately Lebesque-measurable on $[0,1]$, separately continuous on $\mathbf{R}^d \times \mathbf{R}^d$, $||h(t,x,y)|| \leq \alpha(t)$, $\forall (t,x,y) \in [0,T] \times \mathbf{R}^d \times \mathbf{R}^d$ where α is positive Lebesque-integrable.Then the $W_E^{2,1}([0,1])$-solution set to*

$$\begin{cases} \ddot{u}(t) + \gamma \dot{u}(t) = h(t, u(t), \dot{u}(t)) - \nabla \varphi(u(t)) \text{ a.e. } t \in [0,1] \\ u(0) = x, \quad u(1) = \sum_{i=1}^{m-2} \alpha_i u(\eta_i) \end{cases}$$

is compact in the space $\mathcal{C}_E([0,T])$.

Proof The proof is immediate by applying Proposition 2.3 with $F(t,x,y) = h(t,x,y) - \nabla\varphi(x)$, $\forall (t,x,y) \in [0,1] \times E \times E$ and by observing that the subdifferential $x \mapsto \partial\varphi(x) = \nabla\varphi(x)$ is bounded and continuous. ■

3 Applications. Towards the Variational Convergence in Second Order Evolution Inclusions

Let us recall a useful Gronwall type lemma [12].

Lemma 3.1 (A Gronwall-like inequality) *Let $p, q, r : [0,T] \to [0,\infty[$ be three nonnegative Lebesgue integrable functions such that for almost all $t \in [0,T]$*

$$r(t) \leq p(t) + q(t) \int_0^t r(s)\, ds.$$

Then

$$r(t) \leq p(t) + q(t) \int_0^t [p(s) \exp(\int_s^t q(\tau)\, d\tau)]\, ds$$

for all $t \in [0,T]$.

We recall below some notations and summarize some results which describe the limiting behavior of a bounded sequence in $L_H^1([0,T])$. See [10, Proposition 6.5.17].

Proposition 3.1 *Let H be a separable Hilbert space. Let (ζ_n) be a bounded sequence in $L_H^1([0,T])$. Then the following hold:*

(1) (ζ_n) (up to an extracted subsequence) stably converges to a Young measure ν that is, there exist a subsequence (ζ_n') of (ζ_n) and a Young measure ν belonging to the space of Young measure $\mathcal{Y}([0,T]; H_\sigma)$ with $t \mapsto bar(\nu_t) \in L_H^1([0,T])$ (here $bar(\nu_t)$ denotes the barycenter of ν_t) such that

$$\lim_{n\to\infty} \int_0^T h(t, \zeta_n'(t)))\, dt) = \int_0^T \left[\int_H h(t,x)\, \nu_t(dx) \right] dt$$

for all bounded Carathéodory integrands $h : [0, T] \times H_\sigma \to \mathbf{R}$,

(2) (ζ_n) *(up to an extracted subsequence) weakly biting converges to an integrable function* $f \in L^1_H([0, T])$, *which means that there is a subsequence* (ζ'_m) *of* (ζ_n) *and an increasing sequence of Lebesgue-measurable sets* (A_p) *with* $\lim_p \lambda(A_p) = 1$ *and* $f \in L^1_H([0, T])$ *such that, for each* p,

$$\lim_{m\to\infty} \int_{A_p} \langle h(t), \zeta'_m(t)\rangle\, dt = \int_{A_p} \langle h(t), f(t)\rangle\, dt$$

for all $h \in L^\infty_H([0, T])$,

(3) (ζ_n) *(up to an extracted subsequence) Komlós converges to an integrable function* $g \in L^1_H([0, T])$, *which means that there is a subsequence* $(\zeta_{\beta(m)})$ *and an integrable function* $g \in L^1_H([0, T])$, *such that*

$$\lim_{n\to\infty} \frac{1}{n} \Sigma^n_{j=1} \zeta_{\gamma(j)}(t) = g(t), \text{ a.e. } \in [0, T],$$

for every subsequence $(f_{\gamma(n)})$ *of* $(f_{\beta(n)})$.

(4) *There is a filter* $\mathcal{U}$ *finer than the Fréchet filter such that* $\mathcal{U} - \lim_n \zeta_n = l \in (L^\infty_H)'_{weak}$ *where* $(L^\infty_H)'_{weak}$ *is the second dual of* $L^1_H([0, T])$.
Let $w_{l_a} \in L^1_H([0, T])$ *be the density of the absolutely continuous part* l_a *of* l *in the decomposition* $l = l_a + l_s$ *in absolutely continuous part* l_a *and singular part* l_s.
If we have considered the same extracted subsequence in (1)–(4), then one has

$$f(t) = g(t) = bar(\nu_t) = w_{l_a}(t) \text{ a.e. } t \in [0, T].$$

By $W^{2,1}_{\mathbf{R}^d}([0, T])$ (resp. $W^{2,2}_{\mathbf{R}^d}([0, T])$ we denote the set of all continuous functions in $\mathcal{C}_{\mathbf{R}^d}([0, T])$ such that their first derivatives are continuous and their second derivatives belong to $L^1_{\mathbf{R}^d}([0, T])$ (resp. $L^2_{\mathbf{R}^d}([0, T])$) and by $W^{1,1}_{BV}([0, T])$ we denote the set of all continuous functions in $\mathcal{C}_{\mathbf{R}^d}([0, T])$ such that their first derivatives are of bounded variation (BV for short).

We begin with a preliminary result which shows the limiting properties of $W^{2,1}_{\mathbf{R}^d}([0, 1])$-solutions for a second order ordinary differential equation with m-point boundary conditions.

Proposition 3.2 *Let* $E = \mathbf{R}^d$. *Let* $(f_n)_{n\in\mathbf{N}}$ *be a bounded sequence in* $L^1_E([0, 1])$. *For each* $n \in \mathbf{N}$, *let us consider the* $W^{2,1}_E([0, 1])$*-solution* $u_n : [0, 1] \to E$ *of the equation*

$$\ddot{u}_n(t) + \gamma \dot{u}_n(t) = f_n(t),\ t \in [0, 1];\quad u_n(0) = x,\quad u_n(1) = \sum_{i=1}^{m-2} \alpha_i u_n(\eta_i).$$

Then there exist a subsequence of (u_n) *still denoted by* (u_n), *a* $W^{1,1}_{BV}([0, 1])$*-function* $u : [0, 1] \to E$ *and a Young measure* $\nu \in \mathcal{Y}([0, 1]; E)$ *such that* $t \mapsto bar(\nu_t)$ *belongs to* $L^1_E([0, 1])$ *which satisfy the following conditions:*

(a) $(u_n(.))$ *converges in* $\mathcal{C}_E([0,1])$ *to* $u(.)$ *with* $u(0) = x$, $u(1) = \sum_{i=1}^{m-2} \alpha_i u(\eta_i)$.
(b) $(\dot{u}_n(.))$ *converges in* $L^1_E([0,1])$ *to* $\dot{u}(.)$.
(c) $(\delta_{\ddot{u}_n})$ *stably converges in* $\mathcal{Y}([0,1], E)$ *to* ν.
(d) *Assume further that the negative parts* $\langle u_n, \ddot{u}_n\rangle^-$ *of the functions* $\langle u_n, \ddot{u}_n\rangle$ *are uniformly integrable in* $L^1_{\mathbf{R}}([0,1])$.
Then

$$\liminf_{n\to\infty} \int_0^1 \langle u_n(t), \ddot{u}_n(t)\rangle\, dt \geq \int_0^1 \langle u(t), bar(\nu_t)\rangle\, dt = \int_0^1 \left[\int_E \langle u(t), x\rangle\, \nu_t(dx)\right] dt.$$

Proof Existence and uniqueness of a $W^{2,1}_E([0,1])$-solution for the equation

$$\ddot{u}_n(t) + \gamma \dot{u}_n(t) = f_n(t),\ t \in [0,1];\ u(0) = x, \quad u(1) = \sum_{i=1}^{m-2} \alpha_i u(\eta_i).$$

are ensured by Proposition 2.1 with integral representation formulas

$$\begin{cases} u_n(t) = e_x(t) + \int_0^1 G(t,s) f_n(s) ds,\ t \in [0,1] \\ \dot{u}_n(t) = \dot{e}_x(t) + \int_0^1 \frac{\partial G}{\partial t}(t,s) f_n(s) ds,\ t \in [0,1] \end{cases}$$

where

$$\begin{cases} e_x(t) = x + A(1 - \sum_{i=1}^{m-2} \alpha_i)(1 - \exp(-\gamma t))x \\ \dot{e}_x(t) = \gamma A \left(1 - \sum_{i=1}^{m-2} \alpha_i\right) \exp(-\gamma t)x \\ A \quad = \left(\sum_{i=1}^{m-2} \alpha_i - 1 + \exp(-\gamma) - \sum_{i=1}^{m-2} \alpha_i \exp(-\gamma(\eta_i))\right)^{-1}. \end{cases}$$

Since $(f_n(.))$ is bounded in $L^1_E([0,1])$ by assumption, $(\dot{u}_n(.))$ is uniformly bounded by using the integral formula for $\dot{u}_n$ and the boundedness of the Green function G given in Lemma 3.1. So $(\dot{u}_n(.))$ is uniformly bounded and bounded in variation. In view of the Helly–Banach theorem (see e.g. [20, p. 11]), we may, by extracting a subsequence, assume that $(\dot{u}_n(.))$ pointwise converges to a BV function $v(.)$. Let us set $u(t) = \int_0^t v(s)\, ds$ for all $t \in [0,1]$. Then $u \in W^{1,1}_{BV}([0,1])$ with $\dot{u}(t) = v(t)$ for almost every $t \in [0,1]$. Then $(\dot{u}_n(.))$ is uniformly bounded and pointwise converges to $v(.)$. By Lebesgue's theorem, we conclude that $(\dot{u}_n(.))$ converges in $L^1_E([0,1])$ to $\dot{u}(.)$. Hence $(u_n(.))$ converges uniformly to $u(.)$ with $u(0) = x$, $u(1) = \sum_{i=1}^{m-2} \alpha_i u(\eta_i)$. It remains to check (c) and (d). Since $(\ddot{u}_n(.))$ is bounded, in view of Proposition 3.1, we may assume that the sequence $(\delta_{\ddot{u}_n})$ of associated Young measures stably converges in $\mathcal{Y}([0,1], E)$ to a Young measure ν such that $t \mapsto \text{bar}(\nu_t)$ belongs to $L^1_E([0,1])$. Let us prove the last Fatou property (d). We may suppose that

$$a := \lim_{n\to\infty} \int_0^1 \langle u_n(t), \ddot{u}_n(t)\rangle\, dt \in \mathbf{R}.$$

Furthermore, since $(\ddot{u}_n(.))$ is bounded in $L^1_E([0,1])$, in view of Proposition 3.1 we may suppose that $(\ddot{u}_n(.))$ weakly biting converges to a function $f \in L^1_E([0,1])$, that is, there exist a subsequence (still denoted by $(\ddot{u}_n(.))$) of $(\ddot{u}_n(.))$ and an increasing sequence of measurable sets (A_p) in $[0,1]$ such that $\lim_{p\to\infty} \lambda(A_p) = 1$, and such that, for each p and for each $g \in L^\infty_E(A_p, A_p \cap \mathcal{L}([0,1]), \lambda|_{A_p})$, the following holds:

$$\lim_{n\to\infty} \int_{A_p} \langle \ddot{u}_n(t), g(t)\rangle\, dt = \int_{A_p} \langle f(t), g(t)\rangle\, dt.$$

Let $\varepsilon > 0$ be given. Pick $N \in \mathbf{N}$ such that

$$\int_{A_N} \langle u(t), f(t)\rangle\, dt \geq \int_{[0,1]} \langle u(t), f(t)\rangle\, dt - \varepsilon,$$

and that

$$\limsup_{n\to\infty} \int_{[0,1]\setminus A_N} \langle u_n(t), \ddot{u}_n(t)\rangle^-\, dt \leq \varepsilon$$

(this is possible because $(\langle u_n, \ddot{u}_n\rangle^-)_n$ is uniformly integrable by hypothesis). As $||u_n(.) - u(.)|| \to 0$ uniformly, it is easy to see that

$$\lim_{n\to\infty} \int_{A_N} ||u_n(t) - u(t)||\; ||\ddot{u}_n(t)||\, dt = 0.$$

See [6, 16] for a more general case. Whence

$$\lim_{n\to\infty} \Big[\int_{A_N} \langle u_n(t), \ddot{u}_n(t)\rangle\, dt - \int_{A_N} \langle u(t), \ddot{u}_n(t)\rangle\, dt\Big] = 0.$$

An easy computation gives

$$\begin{aligned} a &\geq \lim_{n\to\infty} \int_{A_N} \langle u_n(t), \ddot{u}_n(t)\rangle - \limsup_{n\to\infty} \int_{[0,1]\setminus A_N} \langle u_n(t), \ddot{u}_n(t)\rangle^-\, dt \\ &\geq \lim_{n\to\infty} \int_{A_N} \langle u_n(t), \ddot{u}_n(t)\rangle\, dt - \varepsilon. \end{aligned}$$

Finally we get

$$\begin{aligned} a &\geq \lim_{n\to\infty} \int_{A_N} \langle u_n(t), \ddot{u}_n(t)\rangle\, dt - \varepsilon \\ &= \lim_{n\to\infty} \int_{A_N} \langle u(t), \ddot{u}_n(t)\rangle\, dt - \varepsilon \\ &= \int_{A_N} \langle u(t), f(t)\rangle\, dt - \varepsilon \end{aligned}$$

$$\geq \int_{[0,1]} \langle u(t), f(t) \rangle \, dt - 2\varepsilon.$$

By virtue of Proposition 3.1 $f(t) = \text{bar}(\nu_t)$ a.e. The proof is therefore complete because

$$\int_0^1 \langle u(t), \text{bar}(\nu_t) \rangle \, dt = \int_0^1 \left[\int_E \langle u(t), x \rangle \, \nu_t(dx) \right] dt. \qquad \blacksquare$$

The above techniques can be used to prove the existence of a solution for second order evolution inclusion with boundary conditions governed by subdifferential operators of the form

$$f(t) \in \ddot{u}(t) + Mu(t) + \partial\varphi(u(t)), t \in [0, T] \tag{I}$$

where M is positive, φ is a proper convex proper lower semicontinuous function defined on $\mathbf{R}^d$, and $\partial\varphi(u(t))$ is the subdifferential of the function φ at the point $u(t)$ and the perturbation f belongs to $L^2_{\mathbf{R}^d}([0, T])$. Similar results in this direction are obtained by [1–4, 11].

Now we present a fairly general result for the approximating problem via the epiconvergence approach in a second order evolution problem. The applicability of our abstract results will be exemplified below.

Proposition 3.3 *Assume that $M > 0, \beta \in L^2_{\mathbf{R}^+}([0, T])$. For each $n \in \mathbf{N}$, let $\varphi_n : \mathbf{R}^d \to \mathbf{R}^+$ be a convex, Lipschitz function and let φ_∞ be a nonnegative l.s.c proper function defined on $\mathbf{R}^d$ such that $\varphi_n(x) \leq \varphi_\infty(x)$ for all $n \in \mathbf{N}$ and for all $x \in \mathbf{R}^d$. Let $f^n \in L^2_{\mathbf{R}^d}([0, T])$ such that $||f_n(t)|| \leq \beta(t)$, $\forall n \in \mathbf{N}$, $\forall t \in [0, T]$. For each $n \in \mathbf{N}$, let u^n be a $W^{2,1}_{\mathbf{R}^d}([0, T])$-solution to the problem*

$$\begin{cases} f^n(t) \in \ddot{u}^n(t) + M\dot{u}^n(t) + \partial\varphi_n(u^n(t)), t \in [0, T] \\ u^n(0) = u^n_0; \; \dot{u}^n(0) = \dot{u}^n_0. \end{cases}$$

Assume that

(i) f^n *$\sigma(L^2_{\mathbf{R}^d}, L^2_{\mathbf{R}^d})$-converges to $f^\infty \in L^2_{\mathbf{R}^d}([0, T])$,*
(ii) *φ_n epi-converges to φ_∞,*
(iii) $\lim_n u^n(0) = u^\infty_0 \in$ *dom* φ_∞, $\lim_n \varphi_n(u^n(0)) = \varphi_\infty(u^\infty_0)$, *and* $\lim_n \dot{u}^n(0) = \dot{u}^\infty_0$,
(iv) *There exist $r_0 > 0$ and $x_0 \in \mathbf{R}^d$ such that*

$$\sup_{n\in\mathbf{N}} \sup_{v\in\overline{B}_{L^\infty_{\mathbf{R}^d}([0,T])}} \int_0^T \varphi_\infty(x_0 + r_0 v(t)) < +\infty$$

where $\overline{B}_{L^\infty_{\mathbf{R}^d}([0,T])}$ is the closed unit ball in $L^\infty_{\mathbf{R}^d}([0, T])$.

(a) *Then up to extracted subsequences,* (u^n) *converges uniformly to a* $W^{1,1}_{BV}([0,T])$*-function* u^∞ *and* $(\dot u^n)$ *pointwisely converges to a BV function* v^∞ *with* $v^\infty = \dot u^\infty$*, and* $(\ddot u^n)$ *biting converges to a function* $\zeta^\infty \in L^1_{\mathbf{R}^d}([0,T])$ *so that the limit function* u^∞*,* $\dot u^\infty$ *and the biting limit* ζ^∞ *satisfy the variational inclusion*

$$f^\infty \in \zeta^\infty + M\dot u^\infty + \partial I_{\varphi_\infty}(u^\infty)$$

where $\partial I_{\varphi_\infty}$ *denotes the subdifferential of the convex lower semicontinuous integral functional* I_{φ_∞} *defined on* $L^\infty_{\mathbf{R}^d}([0,T])$

$$I_{\varphi_\infty}(u) := \int_0^T \varphi_\infty(u(t))\,dt, \ \ \forall u \in L^\infty_{\mathbf{R}^d}([0,T]).$$

Furthermore $\lim_n \varphi_n(u^n(t)) = \varphi_\infty(u^\infty(t)) < \infty$ *a.e. and* $\lim_n \int_0^T \varphi_n(u^n(t))\,dt = \int_0^T \varphi_\infty(u^\infty(t))dt$*. Subsequently, the energy estimate holds true almost everywhere* $t \in [0,T]$*,*

$$\varphi_\infty(u^\infty(t)) + \frac{1}{2}||\dot u^\infty(t)||^2 = \varphi_\infty(u^\infty_0)) + \frac{1}{2}||\dot u^\infty_0||^2 - \int_0^t \langle M\dot u^\infty(s), \dot u^\infty(s)\rangle ds + \int_0^t \langle \dot u^\infty(s), f^\infty(s)\rangle ds.$$

Further $(\ddot u^n)$ *weakly converges to the vector measure* $m \in \mathcal{M}^b_{\mathbf{R}^d}([0,T])$ *so that the limit functions* $u^\infty(.)$ *and the limit measure* m *satisfy the following variational inequality:*

$$\int_0^T \varphi_\infty(v(t))\,dt \geq \int_0^1 \varphi_\infty(u^\infty(t))\,dt + \int_0^1 \langle -M\dot u^\infty(t) + f^\infty(t), v(t) - u^\infty(t)\rangle\,dt + \langle -m, v - u^\infty\rangle_{(\mathcal{M}^b_{\mathbf{R}^d}([0,T]), \mathcal{C}_{\mathbf{R}^d}([0,T]))}.$$

In other words, the vector measure $-m + [-M\dot u^\infty + f^\infty]dt$ *belongs to the subdifferential* $\partial J_{\varphi_\infty}(u^\infty)$ *of the convex functional integral* J_{φ_∞} *defined on* $\mathcal{C}_{\mathbf{R}^d}([0,T])$ *by* $J_{\varphi_\infty}(v) = \int_0^1 \varphi_\infty(t, v(t))\,dt$*,* $\forall v \in \mathcal{C}_{\mathbf{R}^d}([0,T])$*.*

(b) *There are a filter* $\mathcal{U}$ *finer than the Fréchet filter,* $l \in L^\infty_{\mathbf{R}^d}([0,T])'$ *such that*

$$\mathcal{U} - \lim_n [f^n - \ddot u^n - M\dot u^n] = l \in L^\infty_{\mathbf{R}^d}([0,T])'_{weak}$$

where $L^\infty_{\mathbf{R}^d}([0,T])'_{weak}$ *is the second dual of* $L^1_{\mathbf{R}^d}([0,T])$ *endowed with the topology* $\sigma(L^\infty_{\mathbf{R}^d}([0,T])', L^\infty_{\mathbf{R}^d}([0,T]))$ *and* $\mathbf{n} \in \mathcal{C}_{\mathbf{R}^d}([0,T])'_{weak}$ *such that*

$$\lim_n [f^n - \ddot u^n - M\dot u^n] = \mathbf{n} \in \mathcal{C}_{\mathbf{R}^d}([0,T])'_{weak}$$

where $\mathcal{C}_{\mathbf{R}^d}([0,T])'_{weak}$ denotes the space $\mathcal{C}_{\mathbf{R}^d}([0,T])'$ endowed with the weak topology $\sigma(\mathcal{C}_{\mathbf{R}^d}([0,T])', \mathcal{C}_{\mathbf{R}^d}([0,T]))$. Let l_a be the density of the absolutely continuous part l_a of l in the decomposition $l = l_a + l_s$ in absolutely continuous part l_a and singular part l_s. Then

$$l_a(h) = \int_0^T \langle h(t), f^\infty(t) - \zeta^\infty(t) - M\dot{u}^\infty(t)\rangle dt$$

for all $h \in L^\infty_{\mathbf{R}^d}([0,T])$ so that

$$I^*_{\varphi_\infty}(l) = I_{\varphi^*_\infty}(f^\infty - \zeta^\infty - M\dot{u}^\infty) + \delta^*(l_s, dom\, I_{\varphi_\infty})$$

where φ^*_∞ is the conjugate of φ_∞, $I_{\varphi^*_\infty}$ the integral functional defined on $L^1_{\mathbf{R}^d}([0,T])$ associated with φ^*_∞, $I^*_{\varphi_\infty}$ the conjugate of the integral functional I_{φ_∞}, $dom\, I_{\varphi_\infty} := \{u \in L^\infty_{\mathbf{R}^d}([0,T]) : I_{\varphi_\infty}(u) < \infty\}$ and

$$\langle \mathbf{n}, h\rangle = \int_0^T \langle f^\infty(t) - \zeta^\infty(t) - M\dot{u}^\infty(t), h(t)\rangle dt + \langle \mathbf{n}_s, h\rangle, \quad \forall h \in \mathcal{C}_{\mathbf{R}^d}([0,T]).$$

with $\langle \mathbf{n}_s, h\rangle = l_s(h)$, $\forall h \in \mathcal{C}_{\mathbf{R}^d}([0,T])$. Further $\mathbf{n}$ belongs to the subdifferential $\partial J_{\varphi_\infty}(u^\infty)$ of the convex lower semicontinuous integral functional J_{φ_∞} defined on $\mathcal{C}_{\mathbf{R}^d}([0,T])$

$$J_{\varphi_\infty}(u) := \int_0^T \varphi_\infty(u(t))\, dt, \ \ \forall u \in \mathcal{C}_{\mathbf{R}^d}([0,T]).$$

(c) *Consequently the density $f^\infty - \zeta^\infty - M\dot{u}^\infty$ of the absolutely continuous part n_a*

$$\mathbf{n}_a(h) := \int_0^T \langle f^\infty(t) - \zeta^\infty(t) - M\dot{u}^\infty(t), h(t)\rangle dt, \quad \forall h \in \mathcal{C}_{\mathbf{R}^d}([0,T])$$

satisfies the inclusion

$$f^\infty(t) - \zeta^\infty(t) - M\dot{u}^\infty(t) \in \partial\varphi_\infty(u^\infty(t)), \quad \text{a.e.}$$

and for any nonnegative measure θ on $[0,T]$ with respect to which $\mathbf{n}_s$ is absolutely continuous

$$\int_0^T h_{\varphi^*_\infty}(\frac{d\mathbf{n}_s}{d\theta}(t))d\theta(t) = \int_0^T \langle u^\infty(t), \frac{d\mathbf{n}_s}{d\theta}(t)\rangle d\theta(t)$$

here $h_{\varphi^*_\infty}$ denotes the recession function of φ^*_∞.

Proof Step 1 $||\dot{u}^n(.)||$ and $\varphi_n(u_n(.))$ are uniformly bounded.

Multiplying scalarly the inclusion

$$f^n(t) - \ddot{u}^n(t) - M\dot{u}^n(t) \in \partial\varphi_n(u^n(t))$$

by $\dot{u}^n(t)$ and applying the chain rule theorem [21, Theorem 2] yields

$$\langle \dot{u}^n(t), f^n(t)\rangle - \langle \dot{u}^n(t), \ddot{u}^n(t)\rangle - \langle \dot{u}^n(t), M\dot{u}_n(t)\rangle = \frac{d}{dt}[\varphi_n(u_n(t))]$$

that is,

$$-\langle M\dot{u}^n(t), \dot{u}^n(t)\rangle + \langle \dot{u}^n(t), f^n(t)\rangle = \frac{d}{dt}\left[\varphi_n(u_n(t)) + \frac{1}{2}||\dot{u}^n(t)||^2\right]. \quad (3.3.1)$$

Integrating this equality on $[0, t]$, we get

$$\begin{aligned}
&\varphi_n(u^n(t)) + \frac{1}{2}||\dot{u}^n(t)||^2 \\
&\quad = \varphi_n(u^n(0)) + \frac{1}{2}||\dot{u}^n(0)||^2 \\
&\qquad - \int_0^t \langle M\dot{u}^n(s), \dot{u}^n(s)\rangle ds + \int_0^t \langle \dot{u}^n(s), f^n(s)\rangle ds \\
&\quad \le \varphi_n(u^n(0)) + \frac{1}{2}||\dot{u}^n(0)||^2 \\
&\qquad + M\int_0^t ||\dot{u}^n(s)||^2 ds + ||f^n||_{L^2_{\mathbf{R}^d}([0,T])}\left(\int_0^t ||\dot{u}^n(s)||^2 ds\right)^{\frac{1}{2}} \\
&\quad \le \varphi_n(u^n(0)) + \frac{1}{2}||\dot{u}^n(0)||^2 \\
&\qquad + M\int_0^t ||\dot{u}^n(s)||^2 ds + \frac{1}{2}||f^n||_{L^2_{\mathbf{R}^d}([0,T])}\left(1 + \int_0^t ||\dot{u}^n(s)||^2 ds\right) \\
&\quad \le \varphi_n(u^n(0)) + \frac{1}{2}||\dot{u}^n(0)||^2 \\
&\qquad + M\int_0^t ||\dot{u}^n(s)||^2 ds + \frac{1}{2}||\beta||_{L^2_{\mathbf{R}}([0,T])}\left(1 + \int_0^t ||\dot{u}^n(s)||^2 ds\right).
\end{aligned}$$

Then, from (iii), the preceding estimate and the Gronwall like inequality (Lemma 3.1), it is immediate that

$$\sup_{n\ge1}\sup_{t\in[0,T]} ||\dot{u}^n(t)|| < +\infty \quad \text{and} \quad \sup_{n\ge1}\sup_{t\in[0,T]} \varphi_n(u^n(t)) < +\infty. \quad (3.3.2)$$

Step 2 Estimation of $||\ddot{u}^n(.)||$. As

$$z^n(t) := f^n(t) - \ddot{u}^n(t) - M\dot{u}^n(t) \in \partial\varphi_n(u^n(t))$$

by the subdifferential inequality for convex lower semi continuous functions we have

$$\varphi_n(x) \geq \varphi_n(u^n(t)) + \langle x - u^n(t), z^n(t)\rangle$$

for all $x \in \mathbf{R}^d$. Now let $v \in \overline{B}_{L^\infty_{\mathbf{R}^d}([0,T])}$, the closed unit ball of $L^\infty_{\mathbf{R}^d}[0, T])$. Taking $x = w(t) := x_0 + r_0 v(t)$ in the preceding inequality we get

$$\varphi_n(w(t)) \geq \varphi_n(u^n(t)) + \langle w(t) - u^n(t), z^n(t)\rangle.$$

Integrating the preceding inequality gives

$$\begin{aligned}\int_0^T \langle x_0 + r_0 v(t) - u^n(t), z^n(t)\rangle dt \\ &= \int_0^T \langle x_0 - u^n(t), z^n(t)\rangle dt + r_0 \int_0^T \langle v(t), z^n(t)\rangle dt \\ &\leq \int_0^T \varphi_n(x_0 + r_0 v(t))dt - \int_0^T \varphi_n(u^n(t))dt.\end{aligned}$$

Whence follows

$$\begin{aligned} r_0 \int_0^T \langle v(t), z^n(t)\rangle dt \leq \int_0^T \varphi_n(x_0 + r_0 v(t))dt \\ - \int_0^T \varphi_n(u^n(t))dt - \int_0^T \langle x_0 - u^n(t), z^n(t)\rangle dt.\end{aligned} \tag{3.3.3}$$

We compute the last integral in the preceding inequality. For simplicity, let us set $v^n(t) = u^n(t) - x_0$ for all $t \in [0, T]$. By integration by parts and taking into account (3.3.2), we have

$$\begin{aligned} -\int_0^T \langle x_0 - u^n(t), z^n(t)\rangle dt &= -\int_0^T \langle v^n(t), \ddot{v}^n(t) + M\dot{v}^n(t)\rangle - f^n(t)\rangle dt \\ = &- [\langle v^n(t), \dot{v}^n(t) + Mv^n(t)]_0^T + \int_0^T \langle \dot{v}^n(t), \dot{v}^n(t) + Mv^n(t)\rangle dt + \int_0^T \langle v^n(t), f^n(t)\rangle dt \\ \leq &- \langle v^n(T), \dot{v}^n(T)\rangle + \langle v^n(0), \dot{v}^n(0)\rangle - \langle Mv^n(T), v^n(T)\rangle \\ &+ \langle Mv^n(0), v^n(0)\rangle + \int_0^T ||\dot{v}^n(t)||^2 dt + \int_0^T \langle \dot{v}^n(t), Mv^n(t)\rangle dt + \int_0^T \langle v^n(t), f^n(t)\rangle dt.\end{aligned} \tag{3.3.4}$$

By (3.3.2)–(3.3.4), we get

$$r_0 \int_0^T \langle v(t), z^n(t)\rangle dt \leq \int_0^T \varphi_\infty(x_0 + r_0 v(t))dt + L \tag{3.3.5}$$

for all $v \in \overline{B}_{L^\infty_{\mathbf{R}^d}([0,T])}$, where L is a generic positive constant independent of $n \in \mathbf{N}$. By (iv) and (3.3.5) we conclude that $(z^n = f^n - \ddot{u}^n - M\dot{u}^n)$ is bounded in $L^1_{\mathbf{R}^d}([0, T])$, then so is $(\ddot{u}^n)$. It turns out that the sequence $(\dot{u}^n)$ is uniformly bounded by using (3.3.2) and is bounded in variation. By Helly theorem, we may assume that $(\dot{u}^n)$ pointwisely converges to a BV function $v^\infty : [0, T] \to \mathbf{R}^d$ and the sequence (u^n) converges uniformly to an absolutely continuous function u^∞ with $\dot{u}^\infty = v^\infty$ a.e. At this point, it is clear that $(\dot{u}_n)$ converges in $L^1_{\mathbf{R}^d}([0, T])$ to v^∞, using (3.3.2) and the dominated convergence theorem. Hence $(M\dot{u}^n(.))$ converges in $L^1_{\mathbf{R}^d}([0, T])$ to $Mv^\infty(.)$.

Step 3 Young measure limit and biting limit of $\ddot{u}_n$. As $(\ddot{u}_n)$ is bounded in $L^1_{\mathbf{R}^d}([0, T])$, we may assume that $(\ddot{u}^n)$ stably converges to a Young measure $\nu \in \mathcal{Y}([0, T]); \mathbf{R}^d)$ with $\text{bar}(\nu) : t \mapsto \text{bar}(\nu_t) \in L^1_{\mathbf{R}^d}([0, T])$ (here $\text{bar}(\nu_t)$ denotes the barycenter of ν_t). Further by Proposition 3.1, we may assume that $(\ddot{u}^n)$ biting converges to a function $\zeta^\infty : t \mapsto \text{bar}(\nu_t)$ that is, there exists a decreasing sequence of Lebesgue-measurable sets (B_p) with $\lim_p \lambda(B_p) = 0$ such that the restriction of $(\ddot{u}_n)$ on each B_p^c converges weakly in $L^1_{\mathbf{R}^d}([0, T])$ to ζ^∞. Note that $(M\dot{u}^n)$ converges in $L^1_{\mathbf{R}^d}([0, T])$ to Mv^∞. It follows that the restriction of $(z^n = f^n - \ddot{u}^n - M\dot{u}^n)$ to each B_p^c weakly converges in $L^1_{\mathbf{R}^d}([0, T])$ to $z^\infty := f^\infty - \zeta^\infty - Mv^\infty$, because (f^n) weakly converges in $L^1_{\mathbf{R}^d}([0, T])$ to f^∞, $(M\dot{u}^n)$ converges in $L^1_{\mathbf{R}^d}([0, T])$ to Mv^∞ and $(\ddot{u}^n)$ biting converges to $\zeta^\infty \in L^1_{\mathbf{R}^d}([0, T])$. It follows that

$$\lim_n \int_B \langle -\ddot{u}^n - W^n(t), w(t) - u^n(t) \rangle = \int_B \langle -\text{bar}(\nu_t) - W(t), w(t) - u(t) \rangle dt \tag{3.3.6}$$

for every $B \in B_p^c \cap \mathcal{L}([0, T])$, and for every $w \in L^\infty_{\mathbf{R}^d}([0, T])$, where $W^n(t) = M\dot{u}^n(t) - f^n(t)$ and $W(t) = M\dot{u}^\infty(t) - f^\infty(t)$. Indeed, we note that $(w(t) - u^n(t))$ is a bounded sequence in $L^\infty_{\mathbf{R}^d}([0, 1])$ which pointwisely converges to $w(t) - u^\infty(t)$, it converges uniformly on every uniformly integrable subset of $L^1_{\mathbf{R}^d}([0, T])$ by virtue of a Grothendieck Lemma [16], recalling here that the restriction of $-\ddot{u}^n - W^n$ on each B_p^c is uniformly integrable. Now, since φ_n lower epiconverges to φ_∞, for every Lebesgue-measurable set A in $[0, T]$, by virtue of Corollary 4.7 in [11], we have

$$+\infty > \liminf_n \int_A \varphi_n(u^n(t))dt \geq \int_A \varphi_\infty(u^\infty(t))dt. \tag{3.3.7}$$

Combining (3.3.2)–(3.3.5)–(3.3.6)–(3.3.7) and using the subdifferential inequality

$$\varphi_n(w(t)) \geq \varphi_n(u^n(t)) + \langle -\ddot{u}^n(t) - W^n(t), w(t) - u^n(t) \rangle$$

gives

$$\int_B \varphi_\infty(w(t))\, dt \geq \int_B \varphi_\infty(u^\infty(t))\, dt + \int_B \langle -\text{bar}(\nu_t) - W(t), w(t) - u^\infty(t) \rangle\, dt.$$

This shows that $t \mapsto -\operatorname{bar}(\nu_t) - W(t)$ is a subgradient at the point u^∞ of the convex integral functional I_{φ_∞} restricted to $L^\infty_{\mathbf{R}^d}(B_p^c)$, consequently,

$$-\operatorname{bar}(\nu_t) - W(t) \in \partial\varphi_\infty(u^\infty(t)), \text{ a.e. on} B_p^c.$$

As this inclusion is true on each B_p^c and $B_p^c \uparrow [0, T]$, we conclude that

$$-\operatorname{bar}(\nu_t) - W(t) \in \partial\varphi_\infty(u^\infty(t)), \text{ a.e. on}[0, T].$$

Step 4 Limit measure in $\mathcal{M}^b_{\mathbf{R}^d}([0, T])$ *of* $\ddot{u}^n$. As $(\ddot{u}_n)$ is bounded in $L^1_{\mathbf{R}^d}([0, T])$, we may assume that $(\ddot{u}^n)$ weakly converges to the vector measure $m \in \mathcal{M}^b_{\mathbf{R}^d}([0, T])$ so that the limit functions $u^\infty(.)$ and the limit measure m satisfy the following variational inequality:

$$\int_0^T \varphi_\infty(v(t))\,dt \geq \int_0^1 \varphi_\infty(u^\infty(t))\,dt + \int_0^1 \langle -M\dot{u}^\infty(t) + f^\infty(t), v(t) - u^\infty(t)\rangle\,dt$$
$$+ \langle -m, v - u^\infty\rangle_{(\mathcal{M}^b_E([0,T]),\mathcal{C}_{\mathbf{R}^d}([0,T]))}.$$

In other words, the vector measure $-m + [-M\dot{u}^\infty + f^\infty]\,dt = -m - W.dt$ belongs to the subdifferential $\partial J_{\varphi_\infty}(u^\infty)$ of the convex functional integral J_{f_∞} defined on $\mathcal{C}_{\mathbf{R}^d}([0, T])$ by $J_{\varphi_\infty}(v) = \int_0^1 \varphi_\infty(v(t))\ dt,\ \forall v \in \mathcal{C}_{\mathbf{R}^d}([0, T])$. Indeed, let $w \in \mathcal{C}_{\mathbf{R}^d}([0, T])$. Integrating the subdifferential inequality

$$\varphi_n(w(t)) \geq \varphi_n(u^n(t)) + \langle -\ddot{u}^n(t) - W^n(t), w(t) - u^n(t)\rangle$$

and noting that $\varphi_\infty(w(t)) \geq \varphi_n(w(t))$ gives immediately

$$\int_0^T \varphi_\infty(w(t))dt \geq \int_0^T \varphi_n(w(t))dt$$
$$\geq \int_0^T \varphi_n(u^n(t))dt + \langle -\ddot{u}^n(t) - W^n(t), w(t) - u^n(t)\rangle dt.$$

We note that

$$\lim_n \int_0^T \langle -W^n(t), w(t) - u^n(t)\rangle dt = \int_0^T \langle -W(t), w(t) - u^\infty(t)\rangle dt$$

because $(W^n := M\dot{u}^n - f^n)$ is uniformly integrable, and weakly converges to $W := M\dot{u}^\infty - f^\infty$ and the bounded sequence in $w(t) - u^n(t)$ pointwise converges to $w - u^\infty$ so that it converges uniformly on uniformly integrable subsets by virtue of Grothendieck lemma. Whence follows

$$\int_0^T \varphi_\infty(w(t))dt \geq \int_0^T \varphi_\infty(u^\infty(t))dt + \int_0^T \langle -W(t), w(t) - u^\infty(t)\rangle dt$$
$$+ \langle -m, w - u^\infty\rangle_{(\mathcal{M}^b_{\mathbf{R}^d}([0,T]), \mathcal{C}_{\mathbf{R}^d}([0,T]))},$$

which shows that the vector measure $-m - W.dt$ is a subgradient at the point u^∞ of the of the convex integral functional J_{φ_∞} defined on $\mathcal{C}_{\mathbf{R}^d}([0, T]))$ by $J_{\varphi_\infty}(v) := \int_0^T \varphi_\infty(v(t))dt, \forall v \in \mathcal{C}_{\mathbf{R}^d}([0, T])$.
Step 5 **Claim** $\lim_n \varphi_n(u^n(t)) = \varphi_\infty(u^\infty(t)) < \infty$ a.e. and $\lim_n \int_0^T \varphi_n(u^n(t))dt = \int_0^T \varphi_\infty(u^\infty(t))dt < \infty$, and subsequently, the energy estimate holds for a.e. $t \in [0, T]$:

$$\varphi_\infty(u^\infty(t)) + \frac{1}{2}||\dot{u}^\infty(t)||^2 = \varphi_\infty(u^\infty(0)) + \frac{1}{2}||\dot{u}^\infty(0)||^2$$
$$- \int_0^t \langle M\dot{u}^\infty(s), \dot{u}^\infty(s)\rangle ds + \int_0^t \langle \dot{u}^\infty(s), f^\infty(s)\rangle ds.$$

With the above results and notations, applying the subdifferential inequality

$$\varphi_n(w(t)) \geq \varphi_n(u^n(t)) + \langle -\ddot{u}^n(t) - W^n(t), w(t) - u^n(t)\rangle$$

with $w = u^\infty$, integrating on $[0, T]$, and passing to the limit when n goes to ∞, gives the inequalities

$$\int_B \varphi_\infty(u^\infty(t))dt \geq \liminf_n \int_B \varphi_n(u^n(t))dt$$
$$\geq \int_B \varphi_\infty(u^\infty(t))dt \geq \limsup_n \int_B \varphi_n(u^n(t))dt$$

on $B \in B^c_p \cap \mathcal{L}([0, T])$ so that

$$\lim_n \int_B \varphi_n(u^n(t))dt = \int_B \varphi_\infty(u^\infty(t))dt \tag{3.3.8}$$

on $B \in B^c_p \cap \mathcal{L}([0, T])$. Now, from the chain rule theorem given in Step 1, recall that

$$\langle \dot{u}^n(t), f^n(t)\rangle - \langle \dot{u}^n(t), \ddot{u}^n(t) - M\dot{u}_n(t)\rangle = \frac{d}{dt}[\varphi_n(u_n(t))],$$

that is,

$$\langle \dot{u}^n(t), z^n(t)\rangle = \frac{d}{dt}[\varphi_n(u_n(t))].$$

By the estimate (3.3.2) and the boundedness in $L^1_{\mathbf{R}^d}([0, T])$ of (z^n), it is immediate that $(\frac{d}{dt}[\varphi_n(u_n(t))])$ is bounded in $L^1_{\mathbf{R}}([0, T])$ so that $(\varphi_n(u_n(.))$ is bounded in variation. By Helly theorem, we may assume that $(\varphi_n(u_n(.))$ pointwisely converges to a BV function ψ. By (3.3.2), $(\varphi_n(u_n(.))$ converges in $L^1_{\mathbf{R}}([0,T])$ to ψ. In particular, for every $k \in L^\infty_{\mathbf{R}^+}([0,T])$ we have

$$\lim_{n\to\infty}\int_0^T k(t)\varphi_n(u_n(t))dt = \int_0^T k(t)\psi(t)dt. \tag{3.3.9}$$

Combining (3.3.8) and (3.3.9) yields

$$\int_B \psi(t)\,dt = \lim_{n\to\infty}\int_B \varphi_n(u^n(t))\,dt = \int_B \varphi_\infty(u^\infty(t))\,dt$$

for all $\in B_p^c \cap \mathcal{L}([0,T])$. As this inclusion is true on each B_p^c and $B_p^c \uparrow [0,T]$, we conclude that

$$\psi(t) = \lim_n \varphi_n(u_n(t)) = \varphi_\infty(u^\infty(t)) \text{ a.e.}$$

Hence we get $\lim_n \varphi_n(u_n(t)) = \varphi_\infty(u^\infty(t))$ a.e. Subsequently, using (iii) the passage to the limit when n goes to ∞ in the equation

$$\begin{aligned}\varphi_n(u^n(t)) + \frac{1}{2}||\dot{u}^n(t)||^2 &= \varphi_n(u^n(0)) + \frac{1}{2}||\dot{u}^n(0)||^2\\ &\quad - \int_0^t \langle M\dot{u}^n(s), \dot{u}^n(s)\rangle ds + \int_0^t \langle \dot{u}^n(s), f^n(s)\rangle ds\end{aligned}$$

yields for a.e. $t \in [0,T]$

$$\begin{aligned}\varphi_\infty(u^\infty(t)) + \frac{1}{2}||\dot{u}^\infty(t)||^2 &= \varphi_\infty(u_0^\infty) + \frac{1}{2}||\dot{u}_0^\infty||^2\\ &\quad - \int_0^t \langle M\dot{u}^\infty(s), \dot{u}^\infty(s)\rangle ds + \int_0^t \langle \dot{u}^\infty(s), f^\infty(s)\rangle ds.\end{aligned}$$

Noting that (f^n) is uniformly integrable and $\dot{u}^n$ is uniformly bounded and pointwise converges to $\dot{u}^\infty$, by virtue of Grothendieck lemma [16], it converges uniformly on uniformly integrable (=relatively weakly compact) subsets of $L^1_{\mathbf{R}^d}([0,T])$, so that

$$\lim_n \int_0^t \langle \dot{u}^n(s), f^n(s)\rangle ds = \int_0^t \langle \dot{u}^\infty(s), f^\infty(s)\rangle ds.$$

Step 6 Localization of further limits and final step.
As $(z^n = f^n - \ddot{u}^n - M\dot{u}^n)$ is bounded in $L^1_{\mathbf{R}^d}([0,T])$, in view of Step 3, it is relatively compact in the second dual $L^\infty_{\mathbf{R}^d}([0,T])'$ of $L^1_{\mathbf{R}^d}([0,T])$ endowed with the weak topology $\sigma(L^\infty_{\mathbf{R}^d}([0,T])', L^\infty_{\mathbf{R}^d}([0,T]))$. Furthermore, (z^n) can be viewed as a bounded sequence in $\mathcal{C}_{\mathbf{R}^d}([0,T])'$. Hence there are a filter $\mathcal{U}$ finer than the Fréchet filter, $l \in L^\infty_{\mathbf{R}^d}([0,T])'$ and $\mathbf{n} \in \mathcal{C}_{\mathbf{R}^d}([0,T])'$ such that

$$\mathcal{U} - \lim_n z^n = l \in L^\infty_{\mathbf{R}^d}([0,T])'_{weak} \tag{3.3.10}$$

and

$$\lim_n z^n = \mathbf{n} \in \mathcal{C}_{\mathbf{R}^d}([0,T])'_{weak} \tag{3.3.11}$$

where $L^\infty_{\mathbf{R}^d}([0,T])'_{weak}$ is the second dual of $L^1_{\mathbf{R}^d}([0,T])$ endowed with the topology $\sigma(L^\infty_{\mathbf{R}^d}([0,T])', L^\infty_{\mathbf{R}^d}([0,T]))$ and $\mathcal{C}_{\mathbf{R}^d}([0,T])'_{weak}$ denotes the space $\mathcal{C}_{\mathbf{R}^d}([0,T])'$ endowed with the weak topology $\sigma(\mathcal{C}_{\mathbf{R}^d}([0,T])', \mathcal{C}_{\mathbf{R}^d}([0,T]))$, because $\mathcal{C}_{\mathbf{R}^d}([0,T])$ is a separable Banach space for the norm sup, so that we may assume by extracting subsequences that (z^n) weakly converges to $\mathbf{n} \in \mathcal{C}_{\mathbf{R}^d}([0,T])'$. Using Step 4, we note that $\mathbf{n} = -m - W.dt = -m - (M\dot{u}^\infty - f^\infty).dt$. Let l_a be the density of the absolutely continuous part l_a of l in the decomposition $l = l_a + l_s$ in absolutely continuous part l_a and singular part l_s, in the sense there is an decreasing sequence (A_n) of Lebesgue measurable sets in $[0,T]$ with $A_n \downarrow \emptyset$ such that $l_s(h) = l_s(1_{A_n}h)$ for all $h \in L^\infty_{\mathbf{R}^d}([0,T])$ and for all $n \geq 1$. As $(z^n = f^n - \ddot{u}^n - M\dot{u}^n)$ biting converges to $z^\infty = f^\infty - \zeta^\infty - M\dot{u}^\infty$ in Step 4, it is already seen (cf. Proposition 3.1) that

$$l_a(h) = \int_0^T \langle h(t), f^\infty(t) - \zeta^\infty(t) - M\dot{u}^\infty(t)\rangle dt$$

for all $h \in L^\infty_{\mathbf{R}^d}([0,T])$, shortly $z^\infty = f^\infty - \zeta^\infty - M\dot{u}^\infty$ coincides a.e. with the density of the absolutely continuous part l_a. By [13, 23], we have

$$I^*_{\varphi_\infty}(l) = I_{\varphi^*_\infty}(f^\infty - \zeta^\infty - M\dot{u}^\infty) + \delta^*(l_s, \operatorname{dom} I_{\varphi_\infty}),$$

where φ^*_∞ is the conjugate of φ_∞, $I_{\varphi^*_\infty}$ is the integral functional defined on $L^1_{\mathbf{R}^d}([0,T])$ associated with φ^*_∞, $I^*_{\varphi_\infty}$ is the conjugate of the integral functional I_{φ_∞} and

$$\operatorname{dom} I_{\varphi_\infty} := \{u \in L^\infty_{\mathbf{R}^d}([0,T]) : I_{\varphi_\infty}(u) < \infty\}.$$

Using the inclusion

$$z^\infty = f^\infty - \zeta^\infty - M\dot{u}^\infty \in \partial I_{\varphi_\infty}(u^\infty),$$

that is,

$$I_{\varphi^*_\infty}(f^\infty - \zeta^\infty - M\dot{u}^\infty) = \langle f^\infty - \zeta^\infty - M\dot{u}^\infty, u^\infty\rangle - I_{\varphi_\infty}(u^\infty),$$

we see that

$$I^*_{\varphi_\infty}(l) = \langle f^\infty - \zeta^\infty - M\dot{u}^\infty, u^\infty\rangle - I_{\varphi_\infty}(u^\infty) + \delta^*(l_s, \operatorname{dom} I_{\varphi_\infty}).$$

Coming back to the inclusion $z^n(t) \in \partial\varphi_n(u^n(t))$, we have

$$\varphi_n(x) \geq \varphi_n(u^n(t)) + \langle x - u^n(t), z^n(t)\rangle$$

for all $x \in \mathbf{R}^d$. By substituting x by $h(t)$ in this inequality, where $h \in L^\infty_{\mathbf{R}^d}([0, T])$, and by integrating

$$\int_0^T \varphi_n(h(t))\, dt \geq \int_0^T \varphi_n(u^n(t))\, dt + \int_0^T \langle h(t) - u^n(t), z^n(t)\rangle\, dt.$$

Arguing as in Step 4 by passing to the limit in the preceding inequality, involving the epiliminf property for integral functionals $\int_0^T \varphi_n(h(t))dt$ defined on $L^\infty_{\mathbf{R}^d}([0, T])$, it is easy to see that

$$\int_0^T \varphi_\infty(h(t))\, dt \geq \int_0^T \varphi_\infty(u^\infty(t))\, dt + \langle h - u^\infty, \mathbf{n}\rangle.$$

Since this holds, in particular, when $h \in \mathcal{C}_{\mathbf{R}^d}([0, T])$, we conclude that $\mathbf{n}$ belongs to the subdifferential $\partial J_{\varphi_\infty}(u^\infty)$ of the convex lower semicontinuous integral functional J_{φ_∞} defined on $\mathcal{C}_{\mathbf{R}^d}([0, T])$

$$J_{\varphi_\infty}(u) := \int_0^T \varphi_\infty(u(t))\, dt, \quad \forall u \in \mathcal{C}_{\mathbf{R}^d}([0, T]).$$

Now, let $B : \mathcal{C}_{\mathbf{R}^d}([0, T]) \to L^\infty_{\mathbf{R}^d}([0, T])$ be the continuous injection, and let $B^* : L^\infty_{\mathbf{R}^d}([0, T])' \to \mathcal{C}_{\mathbf{R}^d}([0, T])'$ be the adjoint of B given by

$$\langle B^* l, h\rangle = \langle l, Bh\rangle = \langle l, h\rangle, \quad \forall l \in L^\infty_{\mathbf{R}^d}([0, T])', \quad \forall h \in \mathcal{C}_{\mathbf{R}^d}([0, T]).$$

Then we have $B^* l = B^* l_a + B^* l_s$, $l \in L^\infty_{\mathbf{R}^d}([0, T])'$ being the limit of $(z_n = f^n - \ddot{u}^n - M\dot{u}^n)$ under the filter $\mathcal{U}$ given in Sect. 4 and $l = l_a + l_s$ being the decomposition of l in absolutely continuous part l_a and singular part l_s. It follows that

$$\langle B^* l, h\rangle = \langle B^* l_a, h\rangle + \langle B^* l_s, h\rangle = \langle l_a, h\rangle + \langle l_s, h\rangle$$

for all $h \in \mathcal{C}_{\mathbf{R}^d}([0, T])$. But it is already seen that

$$\begin{aligned}\langle l_a, h\rangle &= \langle f^\infty - \zeta^\infty - M\dot{u}^\infty, h\rangle \\ &= \int_0^T \langle f^\infty(t) - \zeta^\infty(t) - M\dot{u}^\infty(t), h(t)\rangle dt, \quad \forall h \in L^\infty_{\mathbf{R}^d}([0, T])\end{aligned}$$

so that the measure $B^* l_a$ is absolutely continuous

$$\langle B^* l_a, h\rangle = \int_0^T \langle f^\infty(t) - \zeta^\infty(t) - M\dot{u}^\infty(t), h(t)\rangle dt, \quad \forall h \in \mathcal{C}_{\mathbf{R}^d}([0, T])$$

and its density $f^\infty - \zeta^\infty - M\dot{u}^\infty$ satisfies the inclusion

$$f^\infty(t) - \zeta^\infty(t) - M\dot{u}^\infty(t) \in \partial\varphi_\infty(u^\infty(t)), \quad \text{a.e.}$$

and the singular part $B^* l_s$ satisfies the equation

$$\langle B^* l_s, h\rangle = \langle l_s, h\rangle, \quad \forall h \in \mathcal{C}_{\mathbf{R}^d}([0, T]).$$

As $B^* l = \mathbf{n}$, using (3.3.10) and (3.3.11), it turns out that $\mathbf{n}$ is the sum of the absolutely continuous measure $\mathbf{n}_a$ with

$$\langle \mathbf{n}_a, h\rangle = \int_0^T \langle f^\infty(t) - \zeta^\infty(t) - M\dot{u}^\infty(t), h(t)\rangle dt, \quad \forall h \in \mathcal{C}_{\mathbf{R}^d}([0, T])$$

and the singular part $\mathbf{n}_s$ given by

$$\langle \mathbf{n}_s, h\rangle = \langle l_s, h\rangle, \quad \forall h \in \mathcal{C}_{\mathbf{R}^d}([0, T]),$$

which satisfies the property: for any nonnegative measure θ on $[0, T]$ with respect to which $\mathbf{n}_s$ is absolutely continuous,

$$\int_0^T h_{\varphi_\infty^*}\left(\frac{d\mathbf{n}_s}{d\theta}(t)\right) d\theta(t) = \int_0^T \langle u^\infty(t), \frac{d\mathbf{n}_s}{d\theta}(t)\rangle d\theta(t),$$

where $h_{\varphi_\infty^*}$ denotes the recession function of φ_∞^*. Indeed, as $\mathbf{n}$ belongs to $\partial J_{\varphi_\infty}(u^\infty)$ by applying Theorem 5 in [23] we have

$$J_{\varphi_\infty}^*(n) = I_{\varphi_\infty^*}\left(\frac{d\mathbf{n}_a}{dt}\right) + \int_0^T h_{\varphi_\infty^*}\left(\frac{d\mathbf{n}_s}{d\theta}(t)\right) d\theta(t) \tag{3.3.12}$$

with

$$I_{\varphi_\infty^*}(v) := \int_0^T \varphi_\infty^*(v(t))dt, \forall v \in L^1_{\mathbf{R}^d}([0, T]).$$

Recall that

$$\frac{d\mathbf{n}_a}{dt} = f^\infty - \zeta^\infty - M\dot{u}^\infty \in \partial I_{\varphi_\infty}(u^\infty),$$

that is,

$$I_{\varphi_\infty^*}\left(\frac{d\mathbf{n}_a}{dt}\right) = \langle f^\infty - \zeta^\infty - M\dot{u}^\infty, u^\infty\rangle_{\langle L^1_{\mathbf{R}^d}([0,T]), L^\infty_{\mathbf{R}^d}([0,T])\rangle} - I_{\varphi_\infty}(u^\infty). \tag{3.3.13}$$

From (3.3.13), we deduce

$$
\begin{aligned}
J^*_{\varphi_\infty}(n) &= \langle u^\infty, \mathbf{n}\rangle_{\langle \mathcal{C}_{\mathbf{R}^d}([0,T]), \mathcal{C}_{\mathbf{R}^d}([0,T])'\rangle} - J_{\varphi_\infty}(u^\infty)\\
&= \langle u^\infty, \mathbf{n}\rangle_{\langle \mathcal{C}_{\mathbf{R}^d}([0,T]), \mathcal{C}_{\mathbf{R}^d}([0,T])'\rangle} - I_{\varphi_\infty}(u^\infty)\\
&= \int_0^T \langle u^\infty(t), f^\infty(t) - \zeta^\infty(t) - M\dot{u}^\infty(t)\rangle dt\\
&\quad + \int_0^T \langle u^\infty(t), \frac{d\mathbf{n}_s}{d\theta}(t)\rangle d\theta(t) - I_{\varphi_\infty}(u^\infty)\\
&= I_{\varphi^*_\infty}\left(\frac{d\mathbf{n}_a}{dt}\right) + \int_0^T \langle u^\infty(t), \frac{d\mathbf{n}_s}{d\theta}(t)\rangle d\theta(t)).
\end{aligned}
$$

Coming back to (3.3.12) we get the equality

$$
\int_0^T h_{\varphi^*_\infty}\left(\frac{d\mathbf{n}_s}{d\theta}(t)\right) d\theta(t) = \int_0^T \langle u^\infty(t), \frac{d\mathbf{n}_s}{d\theta}(t)\rangle d\theta(t)). \qquad \blacksquare
$$

Actually, Proposition 3.3 completes Proposition 4.6 in [7], which is a precursor of some results we present here.

We begin with a second order evolution equation with m-point boundary condition

Proposition 3.4 *Assume that* $E = \mathbf{R}^d$, $M > 0$, $\beta \in L^2_{\mathbf{R}^+}([0,T])$. *For each* $n \in \mathbf{N}$, *let* $\varphi_n : \mathbf{R}^d \to \mathbf{R}^+$ *be a* C^1, *convex, Lipschitz function and let* φ_∞ *be a nonnegative l.s.c proper function defined on* $\mathbf{R}^d$ *such that* $\varphi_n(x) \le \varphi_\infty(x)$ *for all* $n \in \mathbf{N}$ *and for all* $x \in \mathbf{R}^d$. *Let* $f : [0,T] \times E \times E \to E$ *satisfying*

(1) For each $(x, y) \in E \times E$ *the scalar function* $t \mapsto f(t, x, y)\rangle$ *is Lebesgue measurable,*
(2) For each $t \in [0, 1]$, *function* $f(t, ., .)$ *is continuous on* $E \times E$,
(3) $||f(t, x, y)|| \le \beta(t)$ *for all* $(t, x, y) \in [0, 1] \times E \times E$.
For each $n \in \mathbf{N}$, *let* u^n *be a* $W^{2,1}_{\mathbf{R}^d}([0, 1])$*-solution to the approximating problem*

$$
(\mathcal{P}_n) \begin{cases} f(t, u^n(t), \dot{u}^n(t)) = \ddot{u}^n(t) + M\dot{u}^n(t) + \nabla\varphi_n(u^n(t)), t \in [0, 1] \\ u^n(0) = x \in dom\ \varphi_\infty, \quad u_n(1) = \sum_{i=1}^{m-2} \alpha_i u_n(\eta_i) \end{cases}
$$

Assume that

(i) φ_n *epi-converges to* φ_∞,
(ii) $\lim_n \dot{u}^n(0) = \dot{u}^\infty_0$,
(iii) There exist $r_0 > 0$ *and* $x_0 \in \mathbf{R}^d$ *such that*

$$
\sup_{v \in \overline{B}_{L^\infty_{\mathbf{R}^d}([0,1])}} \int_0^T \varphi_\infty(x_0 + r_0 v(t)) < +\infty
$$

where $\overline{B}_{L^\infty_{\mathbf{R}^d}([0,1])}$ *is the closed unit ball in* $L^\infty_{\mathbf{R}^d}([0, 1])$.

(*a*) *Then, up to extracted subsequences,* (u^n) *converges uniformly to a* $W^{1,1}_{BV}$ $([0,1])$*-function* u^∞ *with* $u^\infty(0) = x \in dom\ \varphi_\infty$, $u^\infty(1) = \sum_{i=1}^{m-2} \alpha_i u^\infty(\eta_i)$ *and* $(\dot{u}^n)$ *pointwisely converges to a BV function* v^∞ *with* $v^\infty = \dot{u}^\infty$, *and* $(\ddot{u}^n)$ *biting converges to a function* $\zeta^\infty \in L^1_{\mathbf{R}^d}([0,1])$ *so that the limit function* u^∞, $\dot{u}^\infty$ *and the biting limit* ζ^∞ *satisfy the variational inclusion*

$$(\mathcal{P}_\infty) \qquad f^\infty \in \zeta^\infty + M\dot{u}^\infty + \partial I_{\varphi_\infty}(u^\infty)$$

where $f^\infty(t) := f(t, u^\infty(t), \dot{u}^\infty(t), \forall t \in [0,1]$, $\partial I_{\varphi_\infty}$ *denotes the subdifferential of the convex lower semicontinuous integral functional* I_{φ_∞} *defined on* $L^\infty_{\mathbf{R}^d}([0,1])$ *by*

$$I_{\varphi_\infty}(u) := \int_0^1 \varphi_\infty(u(t))\,dt, \ \ \forall u \in L^\infty_{\mathbf{R}^d}([0,1]).$$

(*b*) $(\ddot{u}^n)$ *weakly converges to the vector measure* $m \in \mathcal{M}^b_E([0,1])$ *so that the limit functions* $u^\infty(.)$ *and the limit measure m satisfy the following variational inequality:*

$$\int_0^1 \varphi_\infty(v(t))\,dt \geq \int_0^1 \varphi_\infty(u^\infty(t))\,dt + \int_0^1 \langle -M\dot{u}^\infty(t) + f^\infty(t), v(t) - u^\infty(t)\rangle\,dt$$
$$+ \langle -m, v - u^\infty\rangle_{(\mathcal{M}^b_{\mathbf{R}^d}([0,1]), \mathcal{C}_E([0,1]))}.$$

(*c*) *Furthermore* $\lim_n \int_0^1 \varphi_n(u^n(t))dt = \int_0^T \varphi_\infty(u^\infty(t))dt.$ *Subsequently the energy estimate*

$$\varphi_\infty\left(u^\infty(t)) + \frac{1}{2}||\dot{u}^\infty(t)||^2 \leq \varphi_\infty(x) + \frac{1}{2}||\dot{u}^\infty_0\right)||^2$$
$$- \int_0^t \langle M\dot{u}^\infty(s), \dot{u}^\infty(s)\rangle ds + \int_0^t \langle \dot{u}^\infty(s), f^\infty(s)\rangle ds$$

holds a.e.

(*d*) *There are a filter* $\mathcal{U}$ *finer than the Fréchet filter,* $l \in L^\infty_{\mathbf{R}^d}([0,1])'$ *such that*

$$\mathcal{U} - \lim_n [f^n - \ddot{u}^n - M\dot{u}^n] = l \in L^\infty_{\mathbf{R}^d}([0,1])'_{weak}$$

where $L^\infty_{\mathbf{R}^d}([0,1])'_{weak}$ *is the second dual of* $L^1_{\mathbf{R}^d}([0,1])$ *endowed with the topology* $\sigma(L^\infty_{\mathbf{R}^d}([0,1])', L^\infty_{\mathbf{R}^d}([0,1]))$ *and* $\mathbf{n} \in \mathcal{C}_{\mathbf{R}^d}([0,1])'_{weak}$ *such that*

$$\lim_n [f^n - \ddot{u}^n - M\dot{u}^n] = \mathbf{n} \in \mathcal{C}_{\mathbf{R}^d}([0,1])'_{weak}$$

where $\mathcal{C}_{\mathbf{R}^d}([0,1])'_{weak}$ *denotes the space* $\mathcal{C}_{\mathbf{R}^d}([0,1])'$ *endowed with the weak topology* $\sigma(\mathcal{C}_{\mathbf{R}^d}([0,1])', \mathcal{C}_{\mathbf{R}^d}([0,1]))$ *so that* $\mathbf{n} = -m - (M\dot{u}^\infty - f^\infty)dt$.

Let l_a be the density of the absolutely continuous part l_a of l in the decomposition $l = l_a + l_s$ in absolutely continuous part l_a and singular part l_s. Then

$$l_a(h) = \int_0^T \langle h(t), f^\infty(t) - \zeta^\infty(t) - M\dot{u}^\infty(t)\rangle dt$$

for all $h \in L^\infty_{\mathbf{R}^d}([0, 1])$ so that

$$I^*_{\varphi_\infty}(l) = I_{\varphi^*_\infty}(f^\infty - \zeta^\infty - M\dot{u}^\infty) + \delta^*(l_s, dom\, I_{\varphi_\infty})$$

where φ^*_∞ is the conjugate of φ_∞, $I_{\varphi^*_\infty}$ the integral functional defined on $L^1_{\mathbf{R}^d}([0, 1])$ associated with φ^*_∞, $I^*_{\varphi_\infty}$ the conjugate of the integral functional I_{φ_∞}, $dom\, I_{\varphi_\infty} := \{u \in L^\infty_{\mathbf{R}^d}([0, 1]) : I_{\varphi_\infty}(u) < \infty\}$ and

$$\langle \mathbf{n}, h\rangle = \int_0^1 \langle f^\infty(t) - \zeta^\infty(t) - M\dot{u}^\infty(t), h(t)\rangle dt + \langle \mathbf{n}_s, h\rangle, \quad \forall h \in \mathcal{C}_{\mathbf{R}^d}([0, 1])$$

with $\langle \mathbf{n}_s, h\rangle = l_s(h)$, $\forall h \in \mathcal{C}_{\mathbf{R}^d}([0, 1])$. Further $\mathbf{n}$ belongs to the subdifferential $\partial J_{\varphi_\infty}(u^\infty)$ of the convex lower semicontinuous integral functional J_{φ_∞} defined on $\mathcal{C}_{\mathbf{R}^d}([0, 1])$

$$J_{\varphi_\infty}(u) := \int_0^1 \varphi_\infty(u(t))\, dt, \ \ \forall u \in \mathcal{C}_{\mathbf{R}^d}([0, 1]).$$

(c) Consequently the density $f^\infty - \zeta^\infty - M\dot{u}^\infty$ of the absolutely continuous part $\mathbf{n}_a$

$$\mathbf{n}_a(h) := \int_0^1 \langle f^\infty(t) - \zeta^\infty(t) - M\dot{u}^\infty(t), h(t)\rangle dt, \quad \forall h \in \mathcal{C}_{\mathbf{R}^d}([0, 1])$$

satisfies the inclusion

$$f^\infty(t) - \zeta^\infty(t) - M\dot{u}^\infty(t) \in \partial\varphi_\infty(u^\infty(t)), \quad \text{a.e.}$$

and for any nonnegative measure θ on $[0, T]$ with respect to which $\mathbf{n}_s$ is absolutely continuous

$$\int_0^1 h_{\varphi^*_\infty}(\frac{d\mathbf{n}_s}{d\theta}(t))d\theta(t) = \int_0^T \langle u^\infty(t), \frac{d\mathbf{n}_s}{d\theta}(t)\rangle d\theta(t)$$

where $h_{\varphi^*_\infty}$ denotes the recession function of φ^*_∞.

Proof Existence of a $W^{2,1}_{\mathbf{R}^d}([0, 1])$-solution for the approximating equation

$$\begin{cases} \ddot{u}_n(t) + M\dot{u}_n(t) + \nabla\varphi_n(u^n(t) = f(t, u^n(t), \dot{u}^n(t)), \ \text{a.e. } t \in [0, 1] \\ u_n(0) = x, \quad u_n(1) = \sum_{i=1}^{m-2} \alpha_i u_n(\eta_i) \end{cases}$$

is ensured by Proposition 2.8 with integral representation formulas

$$\begin{cases} u_n(t) = e_x(t) + \int_0^1 G(t,s)[\ddot{u}_n(t) + M\dot{u}_n(s)]ds, \ t \in [0,1] \\ \dot{u}_n(t) = \dot{e}_x(t) + \int_0^1 \frac{\partial G}{\partial t}(t,s)[\ddot{u}_n(t) + M\dot{u}_n(s)]ds, \ t \in [0,1] \end{cases}$$

$$\begin{cases} e_x(t) = x + A(1 - \sum_{i=1}^{m-2} \alpha_i)(1 - \exp(-\gamma t))x \\ \dot{e}_x(t) = \gamma A \left(1 - \sum_{i=1}^{m-2} \alpha_i\right) \exp(-\gamma t)x \\ A \quad = \left(\sum_{i=1}^{m-2} \alpha_i - 1 + \exp(-\gamma) - \sum_{i=1}^{m-2} \alpha_i \exp(-\gamma(\eta_i))\right)^{-1} \end{cases}$$

where G is the Green function given by Lemma 2.1. Then $u^n(0) = x$ and $u_n(1) = \sum_{i=1}^{m-2} \alpha_i u_n(\eta_i)$.

The rest of the proof follows the same lines as that of Proposition 3.3. ■

The following is a new variant on the existence of solutions for the second order evolution inclusion with m-point boundary condition.

Proposition 3.5 *Let $(\partial\varphi_n)$ $(n \in \mathbf{N} \cup \{\infty\})$ be a sequence of subdifferential operators associated with a sequence of nonnegative normal convex integrands (φ_n) $(n \in \mathbf{N} \cup \{\infty\})$. Assume that the following conditions are satisfied:*

(1) For each $n \in \mathbf{N}$, $|\varphi_n(t,x) - \varphi_n(t,y)| \leq \beta_n(t)||x - y||$ for all $t \in [0,1]$ and for all $x, y \in \mathbf{R}^d$, where β_n is a nonnegative integrable functions.

(2) For each Lebesgue-measurable set $A \in [0,1]$, for each $w \in L^\infty_{\mathbf{R}^d}([0,1])$,

$$\limsup_n \int_A \varphi_n(t, w(t))\, dt \leq \int_A \varphi_\infty(t, w(t))\, dt.$$

(3) For each $t \in [0,1]$, $\varphi_n(t,.)$ lower epiconverges to $\varphi_\infty(t,.)$, that is, for each fixed $t \in [0,1]$, for each (x_n) in $\mathbf{R}^d$, converging to $x \in \mathbf{R}^d$, $\liminf \varphi_n(t,x_n) \geq \varphi_\infty(t,x)$.

For each $n \in \mathbf{N}$, let $u^n : [0,1] \to \mathbf{R}^d$ be a $W^{2,1}_{\mathbf{R}^d}([0,1])$-solution to

$$\begin{cases} \ddot{u}^n(t) + \gamma\dot{u}^n(t) \in \partial\varphi_n(t, u^n(t)), \text{ a.e. } t \in [0,1] \\ u^n(0) = x, \quad u^n(1) = \sum_{i=1}^{m-2} \alpha_i u^n(\eta_i). \end{cases}$$

(4) Assume further that

$$\sup_{n\in\mathbf{N}} \int_0^1 \varphi_n(t, u_n(t))dt < +\infty$$

and

$$\sup_{n\in\mathbf{N}} \int_0^1 |\partial\varphi_n(t, u^n(t))|dt < +\infty.$$

Then the following hold:

(a) *Up to extracted subsequences,* (u^n) *converges uniformly to a* $W^{1,1}_{BV}([0,1])$ *function* u^∞ *with* $u^\infty(0) = x$, $u^\infty(1) = \sum_{i=1}^{m-2} \alpha_i u^\infty(\eta_i)$ *and* $(\dot{u}^n)$ *pointwisely converges to the BV function* $\dot{u}^\infty$, *and* $(\ddot{u}^n)$ *stably converges to a Young measure* $\nu^\infty \in \mathcal{Y}([0,1]; \mathbf{R}^d)$ *with* $t \mapsto bar(\nu_t^\infty) \in L^1_{\mathbf{R}^d}([0,1])$ *(here* $bar(\nu_t^\infty)$ *denotes the barycenter of* ν_t^∞*) such that the limit functions* $u^\infty(.)$, $\dot{u}^\infty(.)$ *and the Young limit measure* ν^∞ *satisfy*

$$\int_0^1 \varphi_\infty(t, u^\infty(t))dt \le \liminf_n \int_0^1 \varphi_n(t, u^n(t))dt$$

consequently

$$\lim_n \int_0^1 \varphi_n(t, u^n(t))dt = \int_0^1 \varphi_\infty(t, u^\infty(t))dt < \infty$$

and

$$bar(\nu_t^\infty) + \gamma \dot{u}^\infty(t) \in \partial\varphi_\infty(t, u^\infty(t)), \text{ a.e.}$$

equivalently the function $t \mapsto bar(\nu_t^\infty) + \gamma\dot{u}^\infty(t)$ *belongs to the subdifferential* $\partial I_{\varphi_\infty}(u^\infty)$ *of the convex lower semicontinuous integral functional* I_{φ_∞} *defined on* $L^\infty_{\mathbf{R}^d}([0,T])$

$$I_{\varphi_\infty}(u) := \int_0^T \varphi_\infty(t, u(t))\, dt, \ \ \forall u \in L^\infty_{\mathbf{R}^d}([0,T]).$$

(b) *Up to extracted subsequences,* (u^n) *converges uniformly to a* $W^{1,1}_{BV}([0,1])$ *function* u^∞ *with* $u^\infty(0) = x$, $u^\infty(1) = \sum_{i=1}^{m-2} \alpha_i u^\infty(\eta_i)$ *and* $(\dot{u}^n)$ *pointwisely converges to the BV function* $\dot{u}^\infty$, $(\ddot{u}^n)$ *weakly converges to* $m^\infty \in \mathcal{M}^b_{\mathbf{R}^d}([0,1])$ *so that the limit functions* $u^\infty(.)$ *and the limit measure* m^∞ *satisfy the variational inequality: for every* $v \in \mathcal{C}_{\mathbf{R}^d}([0,1])$,

$$\int_0^1 \varphi_\infty(t, v(t))\, dt \ge \int_0^1 \varphi_\infty(t, u^\infty(t))\, dt + \int_0^1 \langle \gamma\dot{u}^\infty(t)), v(t) - u^\infty(t)\rangle\, dt \\ + \langle m^\infty, v - u^\infty\rangle_{(\mathcal{M}^b_{\mathbf{R}^d}([0,1]), \mathcal{C}_{\mathbf{R}^d}([0,1]))}.$$

In other words, the vector measure $m^\infty + \gamma\dot{u}^\infty\, dt$ *belongs to the subdifferential* $\partial I_{\varphi_\infty}(u)$ *of the convex functional integral* I_{φ_∞} *defined on* $\mathcal{C}_{\mathbf{R}^d}([0,1])$ *by* $I_{\varphi_\infty}(v) = \int_0^1 \varphi_\infty(t, v(t))\, dt$, $\forall v \in \mathcal{C}_{\mathbf{R}^d}([0,1])$.

Proof Existence of a $W^{2,1}_{\mathbf{R}^d}([0,1])$-solution u^n to

$$\begin{cases} \ddot{u}^n(t) + \gamma\dot{u}^n(t) \in \partial\varphi_n(t, u^n(t)), \text{ a.e. } t \in [0,1] \\ u^n(0) = x, \quad u^n(1) = \sum_{i=1}^{m-2} \alpha_i u^n(\eta_i) \end{cases}$$

is ensured by Proposition 2.7 with integral representation formulas

$$\begin{cases} u^n(t) = e_x(t) + \int_0^1 G(t,s)[\ddot{u}^n(s) + \gamma \dot{u}^n(s)]ds, \ t \in [0,1] \\ \dot{u}^n(t) = \dot{e}_x(t) + \int_0^1 \frac{\partial G}{\partial t}(t,s)[\ddot{u}^n(s) + \gamma \dot{u}^n(s)]ds, \ t \in [0,1] \end{cases}$$

where

$$\begin{cases} e_x(t) = x + A(1 - \sum_{i=1}^{m-2} \alpha_i)(1 - \exp(-\gamma t))x \\ \dot{e}_x(t) = \gamma A \left(1 - \sum_{i=1}^{m-2} \alpha_i\right) \exp(-\gamma t) x \\ A \quad = \left(\sum_{i=1}^{m-2} \alpha_i - 1 + \exp(-\gamma) - \sum_{i=1}^{m-2} \alpha_i \exp(-\gamma(\eta_i))\right)^{-1} \end{cases}$$

where G is the Green function given by Lemma 2.1.
Step 1 (a) As $\sup_n \int_0^1 |\partial\varphi_n(t, u^n(t))|dt < +\infty$, it follows that $(\ddot{u}^n + \gamma\dot{u}^n)$ is bounded in $L^1_{\mathbf{R}^d}([0,1])$, namely

$$\sup_n \int_0^1 ||(\ddot{u}^n(t) + \gamma \dot{u}^n(t)||dt < +\infty,$$

so that, by the representation formulas given above, it is immediate that (u^n) and $(\dot{u}^n)$ are uniformly bounded. Hence $(\ddot{u}^n)$ is bounded in $L^1_{\mathbf{R}^d}([0,1])$ and $(\dot{u}_n(.))$ is bounded in variation because $\sup_n \int_0^1 ||\ddot{u}_n(t)||\,dt < +\infty$. In view of the Helly–Banach theorem, we may, by extracting a subsequence, assume that $(\dot{u}^n(.))$ converges pointwisely to a BV function $v^\infty(.)$. Let us set $u^\infty(t) = \int_0^t v^\infty(s)\,ds$ for all $t \in [0,1]$. Then $u^\infty \in W^{1,1}_{BV}([0,1])$. As $(\dot{u}_n(.))$ is uniformly bounded and pointwise converges to $v^\infty(.)$, by Lebesgue's theorem, we conclude that $(\dot{u}^n(.))$ converges in $L^1_{\mathbf{R}^d}([0,1])$ to $\dot{u}^\infty(.)$. Hence $u^n(.)$ converges uniformly to $u^\infty(.)$ with $u^\infty(0) = x, \quad u^\infty(1) = \sum_{i=1}^{m-2} \alpha_i u^\infty(\eta_i)$. So (a) is proved. From the general compactness result for Young measures, [5, 10] one may assume that $\ddot{u}^n$ stably converge to an Young measure ν^∞. Further, by virtue of Proposition 3.1 we may assume that $(\ddot{u}^n)$ biting converges to the integrable function $\mathrm{bar}(\nu^\infty) : t \mapsto \mathrm{bar}(\nu^\infty_t)$, that is, there exists a decreasing sequence (B_p) of Lebesgue measurable sets with $\lambda(\cap B_p) = 0$ such that the restriction of $(\ddot{u}^n)$ on each B_p^c converges $\sigma(L^1, L^\infty)$ to $\mathrm{bar}(\nu)$. It follows that

$$\lim_n \int_B \langle \ddot{u}^n + \gamma \dot{u}^n(t), w(t) - u^n(t)\rangle\,dt = \int_B \langle \mathrm{bar}(\nu_t) + \gamma \dot{u}^\infty(t), w(t) - u^\infty(t)\rangle\,dt \tag{3.5.1}$$

for every $B \in B_p^c \cap \mathcal{L}([0,1])$, and for every $w \in L^\infty_E([0,1])$ because the sequence $(w - u^n)$ in $L^\infty_{\mathbf{R}^d}([0,1])$ is bounded and pointwise converges to $w - u^\infty$, so it converges uniformly on uniformly integrable subsets of $L^1_{\mathbf{R}^d}([0,1])$. Since (φ_n) lower epiconverges to φ_∞, by Corollary 4.7 in [11], we have

$$\liminf_n \int_A \varphi_n(t, u^n(t))\, dt \geq \int_A \varphi_\infty(t, u^\infty(t))\, dt \tag{3.5.2}$$

for every Lebesgue-measurable set A in $[0, 1]$. Combining (3.5.1), (3.5.2) and Assumption (2), and integrating the subdifferential inequality

$$\varphi_n(t, w(t)) \geq \varphi_n(t, u^n(t)) + \langle \ddot{u}^n(t) + \gamma\dot{u}^n(t), w(t) - u^n(t)\rangle \tag{3.5.3}$$

on each $B \in B_p^c \cap \mathcal{L}([0, 1])$ and for every $w \in L^\infty_{\mathbf{R}^d}([0, 1])$, we get

$$\int_B \varphi_\infty(t, w(t))\, dt \geq \int_B \varphi_\infty(t, u^\infty(t))\, dt + \int_B \langle \mathrm{bar}(\nu_t^\infty) + \gamma\dot{u}^\infty(t), w(t) - u^\infty(t)\rangle\, dt.$$

This shows that $t \mapsto \mathrm{bar}(\nu_t^\infty) + \gamma\dot{u}^\infty(t)$ is a subgradient at the point u^∞ of the convex integral functional I_{φ_∞} restricted to $L^\infty_E(B_p^c)$, consequently,

$$\mathrm{bar}(\nu_t) + \gamma\dot{u}^\infty(t) \in \partial\varphi_\infty(t, u^\infty(t)), \text{ a.e. on } B_p^c.$$

As this inclusion is true on each B_p^c and $B_p^c \uparrow [0, 1]$, we conclude that

$$\mathrm{bar}(\nu_t^\infty) + \gamma\dot{u}^\infty(t) \in \partial\varphi_\infty(t, u^\infty(t)), \text{ a.e. on } [0, 1].$$

Finally, applying the above subdifferential inequality, and putting $w = u^\infty$ in (3.5.3), we deduce

$$\begin{aligned}
&\int_B \varphi_\infty(t, u^\infty(t)dt \\
&\quad \geq \limsup_n \int_B \varphi_n(t, u^\infty(t))dt \\
&\quad \geq \limsup_n \int_B [\varphi_n(t, u^n(t)) + \langle \ddot{u}^n(t) + \gamma\dot{u}^n(t), u^\infty(t) - u^n(t)\rangle]dt \\
&\quad = \limsup_n \int_B \varphi_n(t, u^n(t))dt \geq \liminf_n \int_B \varphi_n(t, u^n(t))dt \\
&\quad \geq \int_B \varphi_\infty(t, u^\infty(t))dt
\end{aligned}$$

because

$$\lim_n \int_B \langle \ddot{u}^n(t) + \gamma\dot{u}^n(t), u^\infty(t) - u^n(t)\rangle]dt = 0$$

recalling that $1_B[\ddot{u}^n + \gamma\dot{u}^n]$ is uniformly integrable. Whence follows

$$\lim_n \int_B \varphi_n(t, u^n(t))dt = \int_B \varphi_\infty(t, u^\infty(t))dt.$$

As this inclusion is true on each B in B_p^c and $B_p^c \uparrow [0, 1]$, we conclude that

$$\lim_n \int_0^1 \varphi_n(t, u^n(t))dt = \int_0^1 \varphi_\infty(t, u^\infty(t))dt.$$

Step 2 (b) Repeating the results in Step 1, up to extracted subsequences, (u^n) converges uniformly to a $W^{1,1}_{BV}([0, 1])$ function u^∞ with $u^\infty(0) = x, u^\infty(1) = \sum_{i=1}^{m-2} \alpha_i u^\infty(\eta_i)$ and $(\dot{u}^n)$ pointwisely converges to the BV function $\dot{u}^\infty$. As $(\ddot{u}_n)$ is L^1-bounded we may assume that $(\ddot{u}_n)$ weakly converges to a vector measure $m^\infty \in \mathcal{M}^b_{\mathbf{R}^d}([0, 1])$ since the Banach space $\mathcal{C}_{\mathbf{R}^d}([0, 1])$ is separable and its topological dual is $\mathcal{M}^b_{\mathbf{R}^d}([0, 1])$. Let $w \in \mathcal{C}_{\mathbf{R}^d}(([0, 1])$. Integrating the subdifferential inequality

$$\varphi_n(t, w(t)) \geq \varphi_n(t, u^n(t)) + \langle \ddot{u}^n(t) + \gamma \dot{u}^n(.), w(t) - u^n(t)\rangle$$

and passing to the limit gives immediately

$$\int_0^1 \varphi_\infty(t, w(t))\, dt \geq \int_0^1 \varphi_\infty(t, u^\infty(t))\, dt + \int_0^1 \langle \gamma \dot{u}^\infty(t), w(t) - u^\infty(t)\rangle\, dt$$
$$+ \langle m^\infty, w - u\rangle_{(\mathcal{M}^b_{\mathbf{R}^d}([0,1]), \mathcal{C}_{\mathbf{R}^d}([0,1]))},$$

which shows that the vector measure $m^\infty + \gamma \dot{u}^\infty\, dt$ belongs to the subdifferential $\partial I_{\varphi_\infty}$ of the convex functional integral I_{φ_∞} defined on $\mathcal{C}_{\mathbf{R}^d}([0, 1])$ by $I_{\varphi_\infty}(v) := \int_0^1 \varphi_\infty(t, v(t))\, dt, \forall v \in \mathcal{C}_{\mathbf{R}^d}([0, 1])$. ■

4 Further Applications: Second Order Evolution Problems with Anti-periodic Boundary Condition

It is worth to focus on the main difference in discussing the various approximating problems.

$$f^n(t) = [\ddot{u}^n(t) + M\dot{u}^n(t)] + \nabla\varphi_n(u^n(t)), t \in [0, T] \tag{4.1}$$
$$f^n(t) \in [\ddot{u}^n(t) + M\dot{u}^n(t)] + \partial\varphi_n(u^n(t)), t \in [0, T] \tag{4.2}$$
$$f^n(t) = -[\ddot{u}^n(t) + M\dot{u}^n(t)] + \nabla\varphi_n(u^n(t)), t \in [0, T] \tag{4.3}$$
$$f^n(t) \in -[\ddot{u}^n(t) + M\dot{u}^n(t)] + \partial\varphi_n(u^n(t)), t \in [0, T]. \tag{4.4}$$

Equations (4.1) and (4.2) are usual in second order dynamical systems. We refer to Attouch et al. [4] and Schatzmann [24] for a deep study of such models. See also the results developed in Propositions 3.2–3.5. Here, according to a traditional vein, we prove the existence of generalized solution with the conservation of energy in (3.3) and (3.4). Meanwhile (4.3) and (4.4) appear in the problem of anti-periodic solution developed in Aizicovici et al. [1–3]. Here in Proposition 4.3 we present a first result of the existence of generalized solution for the problem

$$f(t) \in [\ddot{u}(t) + M\dot{u}(t)] + \partial\varphi(u(t))$$

using the approximating problem (4.2) with application (Proposition 3.4) to problem

$$f(t, u(t), \dot{u}(t)) \in \ddot{u}(t) + M\dot{u}(t) + \partial\varphi(u(t)), t \in [0, T]$$

with m-point boundary condition using the approximating problem

$$f(t, u^n(t), \dot{u}^n(t)) = \ddot{u}^n(t) + M\dot{u}^n(t)] + \nabla\varphi_n(u(t)), t \in [0, T]$$

with m-point boundary condition. Here one can see that the techniques employed in (4.1) and (4.2) cannot be used to develop similar results to (4.3) and (4.4), in particular, we cannot obtain the conservation of energy for the variational limits in (4.3) and (4.4) by contrast with (4.1) and (4.2). So it is worth to mention that our tools allow to study the approximating problem of anti-periodic solution in the framework of Haraux–Okochi with anti-periodic solution

$$\begin{aligned} f^n(t) &= [\ddot{u}^n(t) + M\dot{u}^n(t)] + \nabla\varphi_n(u^n(t)), t \in [0, T], \\ u_n(0) &= -u_n(T). \end{aligned}$$

In our opinion, the general problem of the existence of energy conservation solution to second order evolution inclusion of the form

$$f(t) \in [\ddot{u}(t) + M\dot{u}(t)] + \partial\varphi(u(t)) \tag{4.5}$$

where φ is a lower semicontinuous convex proper function is a difficult problem when the perturbation $f \in L^1_H([0, T])$ and H is a separable Hilbert space.

Now, to finish the paper, we show that our abstract result in Proposition 3.3 and the tool developed therein can be applied to the first order of evolution equation and also to the second order evolution equation with anti-periodic boundary conditions. H. Okochi initiated the study for anti-periodic solutions to evolution equations in Hilbert spaces. Following Okochi's work, A. Haraux proved some existence and uniqueness theorems for anti-periodic solutions by using Brouwer's or Schauder fixed point theorems. Aftabizadeh, Aizicovici and Pavel have studied the anti-periodic solutions to second order evolution equation in Hilbert spaces and Banach spaces by using monotone and accretive operator theory for equations of type (4.3) and (4.4). Here we show the applicability of our abstract result to the study of evolution equations of type (4.1) and (4.2) with anti-periodic boundary condition. For notational convenience let us denote by $\mathcal{H}$ the set of of functions $f \in L^2_{loc}(\mathbf{R}, H)$ such that f is anti-periodic, that is, $f(t + T) = -f(t)$ for all $t \in \mathbf{R}$ and

$$\mathcal{H}_\beta([0,T]) := \{f \in \mathcal{H} : ||f(t)|| \le \beta(t), \beta \in L^2_{\mathbf{R}}([0,T]), t \in [0,T]\}.$$

We begin with some examples in the first order of evolution equation with anti-periodic condition.

Proposition 4.1 *Let $H = \mathbf{R}^d$. Assume that $\varphi_n : \mathbf{R}^d \to [0, +\infty[$ are even, convex, Lipschitz and $\varphi_\infty : \mathbf{R}^d \to [0, +\infty]$ is proper lower semicontinuous convex function such that $\varphi_n(x) \le \varphi_\infty(x)$ for all $n \in \mathbf{N}$ and for all $x \in \mathbf{R}^d$. Let f^n be sequence in $\mathcal{H}_\beta([0,T])$ and let u^n be a $W^{1,2}_{\mathbf{R}^d}([0,T])$-solution to the problem*

$$\begin{cases} f^n(t) \in \dot{u}^n(t) + \partial\varphi_n(u^n(t)) & t \in [0,T], \\ u_n(T) = -u_n(0) \end{cases}$$

Assume that the following conditions are satisfied:

(i) φ_n *epiconverges to* φ_∞,
(ii) $\lim_n u^n(0) = u_0^\infty \in dom\ \varphi_\infty$ *and* $\lim_n \varphi_n(u^n(0)) = \varphi_\infty(u_0^\infty)$.
(iii) f^n $\sigma(L^2_{\mathbf{R}^d}([0,T]), L^2_{\mathbf{R}^d}([0,T]))$*-converges to* $f^\infty \in L^2_{\mathbf{R}^d}([0,T])$.

Then the following hold

(a) Up to extracted subsequences, (u^n) converges pointwisely to an anti-periodic absolutely continuous mapping u^∞ with $u^\infty(T) = -u^\infty(0)$, $(\dot{u}^n)$ $\sigma(L^2_{\mathbf{R}^d}, L^2_{\mathbf{R}^d})$-converges to $\zeta^\infty \in L^2_{\mathbf{R}^d}([0,T])$ with $\zeta^\infty = \dot{u}^\infty$, $\lim_n \varphi_n(u^n(t)) = \varphi_\infty(u^\infty(t)) < +\infty$ a.e. and $\lim_n \int_0^T \varphi_n(u^n(t))dt = \int_0^T \varphi_\infty(u^\infty(t))dt < +\infty$.
(b) $f^\infty - \zeta^\infty \in \partial I_{\varphi_\infty}(u^\infty)$ where $\partial I_{\varphi_\infty}$ denotes the subdifferential of the convex lower semicontinuous integral functional I_{φ_∞} defined on $L^\infty_{\mathbf{R}^d}([0,T])$

$$I_{\varphi_\infty}(u) := \int_0^T \varphi_\infty(u(t))\, dt, \quad \forall u \in L^\infty_{\mathbf{R}^d}([0,T]).$$

Proof Existence of $W^{1,2}_{\mathbf{R}^d}([0,T])$-solution u^n to the problem

$$\begin{cases} f^n(t) \in \dot{u}^n(t) + \partial\varphi_n(u^n(t)) & t \in [0,T], \\ u_n(T) = -u_n(0) \end{cases}$$

is ensured. See Haraux [17], Okochi [22].
Step 1 Estimation of u^n, $\dot{u}^n$, and $\varphi_n(u^n(.)$ Multiplying scalarly the inclusion

$$f^n(t) - \dot{u}^n(t) \in \partial\varphi_n(u^n(t)$$

by $\dot{u}^n(t)$ and applying the chain rule formula [21] for the Lipschitz function φ_n gives

$$\langle \dot{u}^n(t), f^n(t)\rangle - ||\dot{u}^n(t)||^2 = \frac{d}{dt}[\varphi_n(u^n(t))]. \tag{4.1.1}$$

Hence by integration of (4.1.1) on $[0, T]$ and anti-periodicity condition we get the estimate

$$||\dot{u}^n||_{L^2_H([0,T])} \leq ||f^n||_{L^2_H([0,T])}. \tag{4.1.2}$$

From the Poincaré inequality

$$||u^n(t)|| \leq \sqrt{T}\, ||\dot{u}^n||_{L^2_H([0,T])}, \forall t \in [0, T]. \tag{4.1.3}$$

Integrating (4.1.1) on $[0, t]$ we get

$$\begin{aligned} 0 \leq \varphi_n(u^n(t)) &= \varphi_n(u^n(0)) - \int_0^t ||\dot{u}^n(s)||^2 ds + \int_0^t \langle \dot{u}^n(s), f^n(s)\rangle ds \qquad (4.1.4)\\ &\leq \varphi_n(u^n(0)) + \int_0^t \langle \dot{u}^n(s), f^n(s)\rangle ds \end{aligned}$$

so that by using the above estimates (4.1.2)–(4.1.3)–(4.1.4), the weak convergence of f^n in $L^2_H([0, T])$ and (ii) we note that $\varphi_n(u^n(t))$ is uniformly bounded.

Step 2 Using the results in Step 1, up to extracted subsequences (u^n) converges pointwisely to an anti-periodic absolutely continuous mapping u^∞ with $u^\infty(T) = -u^\infty(0)$, $(\dot{u}^n)$ $\sigma(L^2_{\mathbf{R}^d}, L^2_{\mathbf{R}^d})$-converges to $\zeta^\infty \in L^2_{\mathbf{R}^d}([0, T])$ with $\zeta^\infty = \dot{u}^\infty$. For simplicity set $z^n(t) := f^n(t) - \dot{u}^n(t)$. Since we have

$$\langle \dot{u}^n(t), z^n(t)\rangle = \frac{d}{dt}[\varphi_n(u^n(t))]$$

and $\langle \dot{u}^n(.), z^n(.)\rangle$ is bounded in $L^1_{\mathbf{R}}([0, T])$, $\varphi_n(u^n(t))$ is of bounded variation and uniformly bounded.

Claim $\lim_n \varphi_n(u_n(t)) = \varphi_\infty(u_\infty(t)) < \infty$ a.e and $\lim_n \int_0^T \varphi_n(u_n(t))dt = \int_0^T \varphi_\infty(u^\infty(t))dt < \infty$.

From the above estimates and Helly theorem, we may assume that $(\varphi_n(u_n(.))$ pointwisely converges to a BV function θ so that $(\varphi_n(u_n(.))$ converges in $L^1_{\mathbf{R}}([0, T])$ to θ. In particular, for every $k \in L^\infty_{\mathbf{R}^+}([0, T])$, we have

$$\lim_{n\to\infty} \int_0^T k(t)\varphi_n(u_n(t))dt = \int_0^T k(t)\theta(t)dt.$$

Coming back to the inclusion $z^n(t) \in \partial\varphi_n(u^n(t))$, and using the fact that $\varphi_n(x) \leq \varphi_\infty(x)$, $\forall n \in \mathbf{N}$, $\forall x \in \mathbf{R}^d$, we have

$$\varphi_\infty(x) \geq \varphi_n(x) \geq \varphi_n(u^n(t)) + \langle x - u^n(t), z^n(t)\rangle$$

for all $x \in \mathbf{R}^d$. Let $h \in L^\infty_{\mathbf{R}^d}([0, T])$. Substituting x by $h(t)$ in this inequality and by integrating on each measurable set B gives

$$\int_B \varphi_\infty(h(t))\,dt \geq \int_B \varphi_n(h(t))\,dt \geq \int_B \varphi_n(u^n(t))\,dt + \int_B \langle h(t) - u^n(t), z^n(t)\rangle\,dt$$

and passing to the limit in the preceding inequality when n goes to $+\infty$, we get

$$\int_B \varphi_\infty(h(t))\,dt \geq \int_B \theta(t)\,dt + \int_B \langle h(t) - u^\infty(t), z^\infty(t)\rangle\,dt \tag{4.1.5}$$

with $z^\infty = f^\infty - \dot{u}^\infty$. In particular, by taking $h = u^\infty$ we get the estimate

$$\int_B \varphi_\infty(u^\infty(t))\,dt \geq \int_B \theta(t)\,dt$$

for all $B \in \mathcal{L}([0, T])$. By the epi-lower convergence result [11, Corollary 4.7], we have

$$\int_B \theta(t)\,dt = \lim_{n\to\infty} \int_B \varphi_n(u^n(t))\,dt \geq \liminf_{n\to\infty} \int_B \varphi_\infty(u^n(t))\,dt \geq \int_B \varphi_\infty(u^\infty(t))\,dt$$

for all $B \in \mathcal{L}([0, T])$. It turns out that $\varphi_\infty(u^\infty(t)) = \theta(t)$ a.e. and

$$\lim_{n\to\infty} \int_B \varphi_n(u^n(t))\,dt = \int_B \varphi_\infty(u^\infty(t))\,dt < \infty. \tag{4.1.6}$$

From (4.1.5) and (4.1.6) it follows that $f^\infty - \zeta^\infty \in \partial I_{\varphi_\infty}(u^\infty)$ where $\partial I_{\varphi_\infty}$ denotes the subdifferential of the convex lower semicontinuous integral functional I_{φ_∞} defined on $L^\infty_{\mathbf{R}^d}([0, T])$

$$I_{\varphi_\infty}(u) := \int_0^T \varphi_\infty(u(t))\,dt, \ \ \forall u \in L^\infty_{\mathbf{R}^d}([0, T]).$$ ■

Here is a variant of Proposition 4.1.

Proposition 4.2 *Let $H = \mathbf{R}^d$. Assume that $\gamma > 0$, $\varphi_n : \mathbf{R}^d \to [0, +\infty]$ is even, convex, Lipschitz, $\varphi_\infty : \mathbf{R}^d \to [0, +\infty]$ is proper lower semicontinuous convex function such that $\varphi_n(x) \leq \varphi_\infty(x)$ for all $n \in \mathbf{N}$ and for all $x \in \mathbf{R}^d$. Let (f^n) be an anti-periodic sequence in $\mathcal{H}_\beta([0, T])$. Let u^n be a $W^{1,2}_{\mathbf{R}^d}([0, T])$ anti-periodic solution to the problem*

$$\begin{cases} f^n(t) \in \dot{u}^n(t) + \partial\varphi_n(u^n(t)) - \gamma u^n(t),\ t \in [0, T] \\ u_n(T) = -u_n(0). \end{cases}$$

Assume that the following conditions are satisfied:

(i) φ_n epiconverges to φ_∞,
(ii) $\lim_n u^n(0) = u^\infty_0 \in dom\ \varphi_\infty$ and $\lim_n \varphi(u^n(0)) = \varphi_\infty(u^\infty_0)$,
(iii) f^n $\sigma(L^2_{\mathbf{R}^d}([0, T]), L^2_{\mathbf{R}^d}([0, T]))$-converges to $f^\infty \in L^2_{\mathbf{R}^d}([0, T])$.

Then the following hold

(a) *Up to extracted subsequences,* (u^n) *converges pointwisely to an anti-periodic absolutely continuous mapping* u^∞ *with* $u^\infty(T) = -u^\infty(0)$, $(\dot{u}^n)$ $\sigma(L^2_{\mathbf{R}^d}, L^2_{\mathbf{R}^d})$-*converges to* $\zeta^\infty \in L^2_{\mathbf{R}^d}([0,T])$ *with* $\zeta^\infty = \dot{u}^\infty$, $\lim_n \varphi_n(u^n(t)) = \varphi_\infty(u^\infty(t)) < +\infty$ *a.e. and* $\lim_n \int_0^T \varphi_n(u^n(t))dt = \int_0^T \varphi_\infty(u^\infty(t))dt < +\infty$.

(b) $f^\infty - \zeta^\infty \in \partial I_{\varphi_\infty}(u^\infty)$ *where* $\partial I_{\varphi_\infty}$ *denotes the subdifferential of the convex lower semicontinuous integral functional* I_{φ_∞} *defined on* $L^\infty_{\mathbf{R}^d}([0,T])$

$$I_{\varphi_\infty}(u) := \int_0^T \varphi_\infty(u(t))\,dt, \ \ \forall u \in L^\infty_{\mathbf{R}^d}([0,T]).$$

Proof Existence of u^n for the problem

$$\begin{cases} f^n(t) - \dot{u}^n(t) + \gamma u^n(t) \in \partial\varphi_n(u^n(t)) & t \in [0,T], \\ u_n(T) = -u_n(0), \end{cases}$$

is ensured. See Haraux [17], Okochi [22].

Step 1 Estimation of $\dot{u}^n$ *and* u^n. Multiplying scalarly the inclusion

$$f^n(t) - \dot{u}^n(t) + \gamma u^n(t) \in \partial\varphi_n(u^n(t)) \tag{4.2.1}$$

by $\dot{u}^n(t)$ and applying the chain rule formula [21] for the Lipschitz function φ_n gives

$$\langle \dot{u}^n(t), f^n(t)\rangle - ||\dot{u}^n(t)||^2 + \gamma\langle \dot{u}^n(t), u^n(t)\rangle = \frac{d}{dt}[\varphi(u^n(t))]. \tag{4.2.2}$$

Hence by integration in (4.2.1) and anti-periodicity conditions we get the estimate

$$||\dot{u}^n||_{L^2_H([0,T])} \le ||f^n||_{L^2_H([0,T])}. \tag{4.2.3}$$

From the Poincaré inequality,

$$||u^n(t)|| \le \sqrt{T}\,||\dot{u}^n||_{L^2_H([0,T])} \le \sqrt{T}\,||f^n||_{L^2_H([0,T])}. \tag{4.2.4}$$

Integrating (4.2.2), we get

$$0 \le \varphi_n(u^n(t)) = \varphi_n(u^n(0)) - \int_0^t ||\dot{u}^n(s)||^2 ds + \int_0^t \langle \dot{u}^n(s), f^n(s)\rangle ds + \gamma \int_0^t \langle \dot{u}^n(s), u^n(s)\rangle ds$$

We note that

$$\int_0^t \langle \dot{u}^n(s), f^n(s)\rangle ds \le \frac{1}{2}||f^n||_{L^2_H([0,T])}(1 + \int_0^t ||\dot{u}^n(s)||^2 ds) \le \text{Const.}$$

$$\gamma \int_0^t \langle \dot{u}^n(s), u^n(s)\rangle ds \le \text{Const.}||f^n||^2_{L^2_H([0,T])}$$

so that by using the above estimate, the $\sigma(L^2_{\mathbf{R}^d}([0,T]), L^2_{\mathbf{R}^d}([0,T]))$ convergence of f^n and (ii), we conclude that $\varphi_n(u^n(t))$ is uniformly bounded. Now the remainder of the proof is similar to that of Proposition 4.1. ■

We finish the paper with the approximating problem in second order evolution equation with anti-periodic condition

$$\begin{cases} f^n(t) &= \ddot{u}^n(t) + M\dot{u}^n(t) + \nabla\varphi_n(u^n(t)),\\ u^n(T) = -u^n(0). \end{cases}$$

where M is a positive constant, φ_n are convex Lipschitz, C^1, even, functions that epi-converges to a lower semicontinuous convex proper function φ_∞, (f_n) is a sequence in $L^2_H([0,T])$ which weakly converges to a function $f_\infty \in L^2_H([0,T])$. Existence of a $W^{2,2}_{\mathbf{R}^d}([0,T])$ anti-periodic -solution to this approximating problem is well known. See Haraux [17], Okochi [22].

Proposition 4.3 *Let $H = \mathbf{R}^d$, $M \in \mathbf{R}^+$. Assume that $\varphi_n : \mathbf{R}^d \to [0,+\infty[$ is C^1, even, convex, Lipschitz and, $\varphi_\infty : \mathbf{R}^d \to [0,+\infty]$ is proper convex lower semicontinuous with $\varphi_n(x) \le \varphi_\infty(x)$, $\forall x \in \mathbf{R}^d$. Let $f^n \in \mathcal{H}_\beta([0,T])$ Let u^n be a $W^{2,2}_{\mathbf{R}^d}([0,T])$ anti-periodic solution to the approximated problem*

$$\begin{cases} f^n(t) &= \ddot{u}^n(t) + M\dot{u}^n(t) + \nabla\varphi_n(u^n(t)), t \in [0,T],\\ u_n(T) = -u_n(0). \end{cases}$$

Assume that

(i) $f^n \sigma(L^2_H, L^2_H)$ converges to $f^\infty \in L^2_H([0,T])$.
(ii) $\lim_n u^n(0) = u_0^\infty \in dom\ \varphi_\infty$, $\lim_n \varphi_n(u^n(0)) = \varphi_\infty(u_0^\infty)$, and $\lim_n \dot{u}^n(0) = \dot{u}_0^\infty$,
(iii) φ_n epi-converges to φ_∞,
(iv) There exist $r_0 > 0$ and $x_0 \in \mathbf{R}^d$ such that

$$\sup_{v \in \overline{B}_{L^\infty_{\mathbf{R}^d}([0,T])}} \int_0^T \varphi_\infty(x_0 + r_0 v(t))) < +\infty$$

where $\overline{B}_{L^\infty_{\mathbf{R}^d}([0,1])}$ is the closed unit ball in $L^\infty_{\mathbf{R}^d}([0,T])$.

Then the following hold

(a) Up to extracted subsequences, (u^n) converges uniformly to a $W^{1,1}_{BV}([0,T])$ anti-periodic function u^∞ with $u^\infty(T) = -u^\infty(0)$, and $(\dot{u}^n)$ pointwisely converges to

the BV function $\dot{u}^\infty$, *and* $(\ddot{u}^n)$ *biting converges to a function* $\zeta^\infty \in L^1_{\mathbf{R}^d}([0, T])$ *which satisfy the variational inclusion*

$$f^\infty - \zeta^\infty - M\dot{u}^\infty \in \partial I_{\varphi_\infty}(u^\infty)$$

where $\partial I_{\varphi_\infty}$ *denotes the subdifferential of the convex lower semicontinuous integral functional* I_{φ_∞} *defined on* $L^\infty_{\mathbf{R}^d}([0, T])$

$$I_{\varphi_\infty}(u) := \int_0^T \varphi_\infty(u(t))\, dt, \ \ \forall u \in L^\infty_{\mathbf{R}^d}([0, T]).$$

Furthermore

$$\lim_n \varphi_n(u^n(t)) = \varphi_\infty(u^\infty(t)) < \infty \text{ a.e.}$$

$$\lim_n \int_0^T \varphi_n(u^n(t))dt = \int_0^T \varphi_\infty(u^\infty(t))dt < \infty.$$

Subsequently, the estimated energy holds almost everywhere

$$\varphi_\infty(u^\infty(t)) + \frac{1}{2}||\dot{u}^\infty(t)||^2 = \varphi_\infty(u^\infty(0)) + \frac{1}{2}||\dot{u}^\infty(0)||^2 \\ - M\int_0^t ||\dot{u}^\infty(s)||^2\, ds + \int_0^t \langle \dot{u}^\infty(s), f^\infty(s)\rangle ds.$$

Further $(\ddot{u}^n)$ *weakly converges to the vector measure* $m \in \mathcal{M}^b_H([0, T])$ *so that the limit functions* $u^\infty(.)$ *and the limit measure m satisfy the following variational inequality:*

$$\int_0^T \varphi_\infty(v(t))\, dt \geq \int_0^T \varphi_\infty(u^\infty(t))\, dt + \int_0^T \langle -M\dot{u}^\infty(t) + f^\infty(t), v(t) - u^\infty(t)\rangle\, dt \\ + \langle -m, v - u^\infty\rangle_{(\mathcal{M}^b_E([0,T]), \mathcal{C}_E([0,T]))}.$$

In other words, the vector measure $-m + [-M\dot{u}^\infty + f^\infty]\, dt$ *belongs to the subdifferential* $\partial I_{f_\infty}(u)$ *of the convex functional integral* I_{f_∞} *defined on* $\mathcal{C}_H([0, T])$ *by* $I_{\varphi_\infty}(v) = \int_0^T \varphi_\infty(t, v(t))\, dt$, $\forall v \in \mathcal{C}_H([0, T])$.

Proof Existence of $W^{2,2}_{\mathbf{R}^d}([0, T])$-solution u^n for the approximated problem

$$\begin{cases} f^n(t) = \ddot{u}^n(t) + M\dot{u}^n(t) + \nabla\varphi_n(u^n(t)) & t \in [0, T], \\ u_n(T) = -u_n(0) \end{cases}$$

follows from Haraux [17]. Now we can finish the proof by repeating mutatis mutandis the machinery developed in Proposition 3.3. Therefore our $W^{1,1}_{BV}([0, T])$ anti-periodic limit u^∞ of (u^n) and biting limit ζ^∞ of $(\ddot{u}^n)$ satisfies the inclusion

$$f^\infty(t) - \zeta^\infty(t) - M\dot{u}^\infty(t) \in \partial\varphi_\infty(u^\infty(t))$$

and the energy estimate holds

$$\varphi_\infty(u^\infty(t)) + \frac{1}{2}||\dot{u}^\infty(t)||^2 = \varphi_\infty(u_0^\infty) + \frac{1}{2}||\dot{u}_0^\infty||^2 - M\int_0^t ||\dot{u}^\infty(s)||^2 ds + \int_0^t \langle \dot{u}^\infty(s), f^\infty(s)\rangle ds$$

almost everywhere. ■

References

1. Aftabizadeh AR, Aizicovici S, Pavel NH (1992) On a class of second-order anti-periodic boundary value problems. J Math Anal Appl 171:301–320
2. Aftabizadeh AR, Aizicovici S, Pavel NH (1992) Anti-periodic boundary value problems for higher order differential equations in Hilbert spaces. Nonlinear Anal 18:253–267
3. Aizicovici S, Pavel NH (1991) Anti-periodic solutions to a class of nonlinear differential equations in Hilbert space. J Funct Anal 99:387–408
4. Attouch H, Cabot A, Redont P (2002) The dynamics of elastic shocks via epigraphical regularization of a differential inclusion. Barrier and penalty approximations, Advances in mathematical sciences and applications, vol 12, no 1, Gakkotosho, Tokyo, pp 273–306
5. Balder EJ (1997) Lectures on Young measure theory and its applications in economics. Rend Istit Mat Univ Trieste, vol. 31, suppl.:1–69. Workshop di Teoria della Misura et Analisi Reale Grado, 1997 (Italia)
6. Castaing, C (1980) Topologie de la convergence uniforme sur les parties uniformément intégrables de L^1_E et théorèmes de compacité faible dans certains espaces du type Köthe-Orlicz. *Travaux Sém. Anal. Convexe*, 10(1):exp. no. 5, 27
7. Castaing C, Ibrahim AG, Yarou M (2005) Existence problem of second order evolution inclusions: discretization and variational problems. Taiwan J Math 12(6):1435–1477
8. Castaing C, Le Xuan T (2011) Second order differential inclusion with m-point boundary condition. J Nonlinear Convex Anal 12(2):199–224
9. Castaing C, Godet-Thobie Ch, Le Xuan T, Satco B (2014) Optimal control problems governed by a second order ordinary differential equation with m-point boundary condition. Adv Math Econ 18:1–59
10. Castaing C, Raynaud de Fitte P, Valadier M (2004) Young measures on topological spaces. With applications in control theory and probability theory. Kluwer Academic Publishers, Dordrecht
11. Castaing C, Raynaud de Fitte P, Salvadori A (2006) Some variational convergence results with application to evolution inclusions. Adv Math Econ 8:33–73
12. Castaing C, Salvadori A, Thibault L (2001) Functional evolution equations governed by nonconvex sweeping process. J. Nonlinear Convex Anal. 2(2):217–241
13. Castaing C, Valadier M (1977) Convex analysis and measurable multifunctions. Lectures notes in mathematics, vol 580. Springer, Berlin
14. J.P.R Christensen, Topology and borel structure, Math Stud 10. Notas de Mathematica
15. Clarke F (1975) Generalized gradients and applications, Trans Am Math Soc **205**
16. Grothendieck A (1964) Espaces Vectoriels Topologiques, 3rd edn. Publ Soc Mat, São Paulo
17. Haraux A (1989) Anti-periodic solutions of some nonlinear evolution equations. Manuscr Math. 63:479–505
18. Mabrouk M (2003) A variational principle for a nonlinear differential equation of second order. Adv Appl Math 31(2):388–419

19. Maruo K (1986) On certain nonlinear differential equation of second order in time. Osaka J Math 23:1–53
20. Monteiro Marques MDP (1993) Differential inclusions non nonsmooth mechanical problems, shocks and dry friction. Progress in nonlinear differential equations and their applications, vol 9. Birkhauser, Basel
21. Moreau JJ, Valadier M (1987) A chain rule involving vector functions of bounded variations. J Funct Anal 74(2):333–345
22. Okochi H (1988) On the existence of periodic solutions to nonlinear abstract parabolic equations. J Math Soc Jpn 40(3):541553
23. Rockafellar RT (1971) Integrals which are convex functionals, II. Pac J Math 39(2):439–469
24. Schatzmann M (1979) Problèmes unilatéraux d'évolution du 2ème ordre en temps. Ph.D. thesis, Université Pierre et Marie Curie, Paris, (Thèse de Doctorat ès–Sciences – Mathématiques)
25. Thibault L (1976) Propriétés des sous-différentiels de fonctions localement Lipschitziennes définies sur un espace de Banach séparable. Applications, Thèse, Université Montpellier

On Sufficiently-Diffused Information in Bayesian Games: A Dialectical Formalization

M. Ali Khan and Yongchao Zhang

Abstract There have been substantive recent advances in the existence theory of pure-strategy Nash equilibria (PSNE) of finite-player Bayesian games with diffused and dispersed information. This work has revolved around the identification of a *saturation* property of the space of information in the formalization of such games. In this paper, we provide a novel perspective on the theory through the extended Lebesgue interval presented in Khan and Zhang (Adv Math 229:1080–1103, 2012) [26] in that (i) it resolves the existing counterexample of Khan–Rath–Sun (J Math Econ 31:341–359, 1999) [17], and yet (ii) allows the manufacture of new examples.

JEL Classification Numbers: C62, D50, D82, G13
2010 Mathematics Subject: 28C99, 28E05, 91A13, 91A44

This work draws on results in a paper titled "On Sufficiently Diffused Information and Finite-Player Games with Private Information," and initially presented at the workshop on *The Probabilistic Impulse behind Modern Economic Theory* held by the Institute for Mathematical Sciences, National University of Singapore (NUS) during January 11–18, 2011. A substantial version formed the basis of a talk by Khan at *The 6th Conference on Mathematical Analysis in Economic Theory* held at Keio University, January 26–29, 2015. He thanks Professor Toru Maruyama for his invitation, and him and Professors Bob Anderson, Damien Eldridge, Josh Epstein, Chiaki Hara, Alexander Ioffe, Takashi Kamihigashi, Nobusumi Sagara, Takashi Suzuki and Vladimir Tikhomirov for delightful and stimulating conversation at the Conference. Both authors also thank Hülya Eraslan, John Quah and Metin Uyanik for their comments and encouragement of this project. Section 4 of the Keio version was presented at a study group on *Statistical Decision Theory* organized by Professors Idione Meneghel and Rabee Tourky when the author held the position of Visiting Research Fellow at the *Australian National University*, February 15-April 15, 2016. This final version is to be presented at a session organized by Professor Jean-Marc Bonnisseau at the *16th Conference of the Society for the Advancement of Economic Theory* to be held at IMPA, July 6–9, 2016. This research program is also supported by NSFC (11201283).

M. Ali Khan (✉)
Department of Economics, The Johns Hopkins University, Baltimore, MD 21218, USA
e-mail: akhan@jhu.edu

Y. Zhang
School of Economics, Shanghai University of Finance and Economics, 777 Guoding Road, Shanghai 200433, China
e-mail: yongchao@mail.shufe.edu.cn

S. Kusuoka and T. Maruyama (eds.), *Advances in Mathematical Economics*, Advances in Mathematical Economics 21,
DOI 10.1007/978-981-10-4145-7_2

Through the formulation of a *d-property* of an abstract probability space, we exhibit a process under which a game without a PSNE in a specific class of games can be upgraded to one with: a (counter)example on any n-fold extension of the Lebesgue interval resolved by its $(n+1)$-fold counterpart. The resulting dialectic that we identify gives insight into both the *saturation* property and its recent generalization proposed by He–Sun–Sun (Modeling infinitely many agents, working paper, National University of Singapore, 2013) [14] and referred to as *nowhere equivalence.* The primary motivation of this self-contained essay is to facilitate the diffusion and use of these ideas in mainstream non-cooperative game theory. (190 words).

Keywords Bayesian games · d-property · Saturation property · KRS-like games · Lebesgue extension · Nowhere equivalent σ-algebras

Article type: Research Article
Received: July 29, 2015
Revised: June 21, 2016

> How Carathéodory came to think of this definition seems mysterious, since it is not in the least intuitive. It is rather difficult to get an understanding of the meaning of ... measurability except through familiarity with its implications... Carathéodory's definition has many useful implications. The greatest justification of this apparently complicated concept is, however, its possibly surprising but absolutely complete success as a tool in proving the important and useful extension theorem.
>
> (Halmos (1950) and Hewitt-Stromberg (1965))[1]

1 Introduction

In two ground-breaking papers published in 1950–1951, Nash defined for a finite game what, in anachronistic hindsight, is now termed a pure-strategy Nash equilibrium (henceforth PSNE) for a classical setting.[2] As is well-understood, Nash could not prove the existence of such an equilibrium for his game-theoretic set-up because of the obvious reason that it was not true in general. Two decades were to pass before Schmeidler [44] presented an existence proof for such an equilibrium in a class of games with a continuum of players, each of whose payoffs were restricted to depend on a suitably-defined aggregate of all the other players' actions, rather than on each individual action as in Nash. In a complete information setting of one-shot simultaneous play, Schmeidler retained the assumption of a finite-action set for each player, and pointed out how his existence theorem, apart from being of interest for its own sake, implies the existence of a mixed-strategy Nash equilibrium, again as defined and shown by Nash in 1950–1951. Indeed, 1973–1974 were significant

[1]The quotations are taken from Nillsen ([37], p. 340). The authors are grateful to Ashvin Rajan for bringing Nillson's book to their attention.

[2]These classical papers are well-known and now collected in [36].

years for non-cooperative game theory as developed by Nash: Harsanyi in 1973 and Aumann in 1974 presented scenarios in which pure-strategies taken under incomplete information can be seen as rationalizing a given mixed-strategy Nash equilibrium of a classical finite game.[3] Harsanyi focused on *disturbed (perturbed)* games, while Aumann considered games with private information and *subjective* beliefs. Both papers used as their subtext Harsanyi's pioneering papers in 1967–1968 on games with incomplete information. We shall not have anything to say on Harsanyi's work; his formulation and results drew on formalizations of genericity, and thereby applied to *almost all* games belonging to a well-specified class as opposed to a given game.[4]

The equivalence theorem of Aumann, on the other hand, deserves to be even better known than it is. Given a mixed-strategy Nash equilibrium of a classical finite player game of complete information, Aumann can be read as posing the question of what conditions on a space of information and of subjective beliefs would guarantee that the given equilibrium can be induced by each player playing a pure-strategy, where the notion of a pure-strategy strategy is now lifted up from being a point in an action set to being a function, a random variable, from the space of information to the action set. In other words, Aumann asked for conditions on information and beliefs that allow an equilibrium probability distribution of a classical finite game to be induced by random variables in equilibrium. As is by now well-known and well-understood, Aumann required that a player's information be independent of, and his beliefs be atomless on, the pooled information of all the other players. Succinctly put, and in the vernacular that was subsequently to follow, it required the space of information be rich enough so as to allow *independent atomless* supplements. Aumann's equivalence theorem is relevant to us here because of the significant role that it has played for the formulation of games with private information.[5]

The literature on games with non-atomic measure spaces then bifurcates into two distinct branches. Schmeidler's paper originates the theory of large one-shot games of complete information in which the existence of pure and mixed strategy equilibria, as well as the relationship between them, as captured by the notion of a *purification,* is investigated. We may also mention here Mas-Colell's complementation of Schmeidler's existence result on non-anonymous (individualized) atomless games by anonymous (distributionalized) ones.[6] In Mas-Colell's setting too, pure and mixed strategy equilibria, as well as the relationship between them, as captured by the notion of a *symmetrization,* is investigated. Radner–Rosenthal [40], henceforth RR,

[3]These papers are now classical and well-known: for Harsanyi's papers, see [10], and for Aumann's, [1]. In particular, we shall make extensive reference to the 1974 paper of Aumann's and the 1983 paper of Aumann et al., and these are respectively Chaps. 31 and 30 in [1].

[4]Note that the 1973 paper of Harsanyi's is available in [10, Sect. B], and so we only reference the latter in our bibliography; also see Footnote 3 above. For games with incomplete information, see [10, Sect. 1]; and also Myerson [35].

[5]The authors have revisited Aumann's equivalence theorem in [29], and the reader should not confuse it with the core equivalence theorem. It is important for the record to note that this was the background paper at Khan's talk at Tokyo.

[6]See Khan et al. [21, 22] for the terms large *individualized* and large *distributionalized* games, LIG and LDG respectively, and references to the antecedent literature on the concepts they name.

is the originating paper for the other branch. It is rooted in Aumann's equivalence theorem and is, in its undiluted essence, simply the observation that there are no independent atomless supplements for an arbitrary measure space of information. In three decisive examples, only the first of which is our concern here, they showed that a two-player matching-pennies game, when converted into a game of private information with each player's information formalized by the Lebesgue interval, and the joint space of information by the lower triangle of the Lebesgue square, has no pure-strategy equilibria in the sense defined by Aumann. As such, there is then no possibility of an equivalence theorem whereby the mixed strategy equilibrium of the given matching-pennies game, one in which each player putting equal probability on each of the two actions, can be induced by a pure-strategy equilibrium of the game with private information. There is no pure-strategy equilibrium in the game underlying the RR example, and so it inducing the given mixed strategy equilibrium of the given matching-pennies game does not arise: it is aborted right at the very beginning.

The RR example proved decisive for Aumann's equivalence theorem. However, if the assumption of independent and atomless information (*dispersed* and *disparate* in the vernacular of RR) was made right at the outset, rather than as an extraction requirement for the given space of information and beliefs, and the game with private information as an object of interest in its own right, the question can be reformulated from the search for an equivalence theorem to that for an existence theorem. Indeed, such an existence theorem is an obvious consequence of Aumann's result, paired with Nash's existence proved. If one allows the additional wrinkle that players' payoffs also depend on their private information, one could show the existence of pure-strategy equilibria for a game of private information. These are the 1982 RR existence results. Their RR paper thus originates the theory of large one-shot games of incomplete information in which the existence of pure and mixed strategy equilibria, as well as the relationship between them, again formalized by the notion of *purification,* is investigated; also see here the contributions of Milgrom–Weber [34] and their followers; also [42, 46]. But it bears emphasis that the resulting theory, in so far as pure-strategy equilibria are concerned, is constrained to finite action sets, just as it is in the Schmeidler–Mas-Colell set up.

The question then arises as to what happens to both branches of this theory of non-atomic non-cooperative games when the restrictive assumption of finite-action sets no longer holds. It took another decade beyond the Radner–Rosenthal–Milgrom–Weber papers for a picture to emerge; see [40]. The outlines of this are by now well-understood, and the details available in the PNAS announcement and the Handbook chapter published as Khan–Sun [24], Khan–Sun [25] respectively. This need not concern us here other than the following summary statement:

(i) Though they require some non-trivial technical work, the results all generalize to denumerably-infinite action sets with arbitrary atomless measure spaces.
(ii) The results do not hold in general for uncountably-infinite action sets with arbitrary atomless measure spaces.
(iii) The results hold for uncountably-infinite action sets if one restricts attention to atomless Loeb measure spaces, as in Loeb [30].

By 2005, it was well-understand that the entire theory could be generalized beyond atomless Loeb spaces to what were termed *saturated* or *super-atomless* measure spaces. The particular name was hardly the issue: the point was that one could work with abstract measures spaces of a type identified in Maharam [33], and conveniently characterized by Hoover–Keisler [11], and that these spaces were not only sufficient for the existence results, as atomless Loeb spaces were, but also necessary in some well-specified sense. The actual publications originating this new direction were Carmona–Podczeck [4] and Keisler–Sun [16].[7] The point is that the σ-algebra of a saturated or a super-atomless measure spaces is one which, modulo null sets, is nowhere countably-generated. But what is it really? And what does it mean to say that it is necessary? This essay is devoted to the pursuit of an answer to this question. However, it is important for the reader to understand that in the sequel, we do this only in the context of the use of these measure spaces in finite Bayesian games with private information – the treatment of large games with complete information does not concern us here.

A saturated probability space is in some sense an idealized limit space, but to repeat, what is this sense? Again, even though one grants the validity of the necessity result, as one must, the question nags as the substantive meaning of this claim. What does it mean to say that a saturated space is necessary for the existence question? And why is this result of any substantive (economic or game-theoretic) importance? To be sure, the mathematical definition of a saturated space, and the various equivalences underlying it, are clear enough,[8] but what is its meaning in terms of the language and vocabulary that mathematical economists and game theorists are used to? and also what is its characterization in terms of the mathematics with which they work, and are at ease, with? This essay then is addressed, at least in the first instance, to these questions. It takes as its point of departure a neglected 1999 (KRS) example on the non-existence of equilibria in Bayesian games based on an interval as a common action set, and the Lebesgue interval (LI) as the space of private information or types.[9]

In [28], the authors introduced the notion of a *KRS-like game* based on the KRS example, and that of a measure spaces with the *d-property* with respect to a measurable, measure-preserving function and thereby with respect to a sub-σ-algebra. These two concepts, though not technically intricate or in themselves mathematically deep, can nevertheless be used to give insight and feel for answers to the questions posed in the paragraph above. Specifically, we use these two concepts as crucial levers to show that:

---

[7]There is some controversy stemming from the fact that the results in Keisler–Sun [16] were obtained in 2002; see their acknowledgement, and also the use of their results by Noguchi in 2008. It is our firm intention not to get bogged down in this controversy here.

[8]See Hoover–Keisler [11], and the comprehensive discussion in Fajardo–Keisler [5]; also the papers of, Carmona–Podczeck [4] and Keisler–Sun [16]. In a recent important paper, Modukhovich–Sagara [32] establish the relevance of the property in stochastic models of dynamic programming.

[9]See Khan–Rath–Sun [17], and also its footnote to the Fudenberg-Tirole text as to the possible reason why it has remained neglected. Note that this example does not invoke any order structures on the action sets.

(i) an equilibrium *does* exist in the KRS example if the information spaces are upgraded from the unit Lebesgue interval (LI) to the extended LI presented in [26],
(ii) there exists an upgraded KRS example of a game without an equilibrium when modeled on the extended LI,
(iii) the upgrading process reveals an infinite recursion in that a (counter)example can always be constructed if the information spaces are modeled on any n-fold extended[10] LI, n a natural number, but one which can be resolved by a $(n+1)$-fold extended LI, (Proposition 2 below),[11]
(iv) this "recursive upgrading" or "dialectic" then suggests the formulation of a *KRS-like game*, one based on an abstract, atomless probability space, for which a characterization and existence of PSNE can be established (Proposition 1 and Theorem 1 below),[12]
(v) this infinite recursion establishes the importance of KRS-like games as a diagnostic tool to check whether a given information structure guarantees the existence of a PSNE for a general class of *all* private information games,
(vi) a visual and analytical content can be imparted into private information structures that are *relatively-diffused,* as in He–Sun [13], or *saturated,* as in [27, 28].[13]

We now turn to an extended outline and overview of this essay.

After presenting the basic model and the antecedent results in Sect. 2, we recall in Sect. 3 the principle result in [28] based on the two notions of a notion of a *KRS-like* games, and the *relative d-property with respect to a measure-preserving map.* KRS-like games are two-player games with the interval $[-1, 1]$ as the (common) action set, arbitrary atomless probability spaces, and with a structure of payoffs that lead their equilibrium distributions, potential or otherwise, to have the same sort of structure as those of the KRS example; see Proposition 1 below.[14] To be sure, one could consider such games modeled on compact metric action spaces,[15] but as we shall see in the sequel, the underlying motivation for such games is to find the simplest setting that illustrates, and can be used as a criterion for getting a handle on, the difficulties that

[10]This notation then would lead the LI to be viewed as 0-fold extended LI and the extended LI in (i) above as a 1-fold extended LI.

[11]We shall be referring to this below as a "scrambling" operation on a particular game.

[12]In particular, the upgraded games in (i) and (ii) above belong to the class of KRS-like games that is being singled out and studied in this paper.

[13]The references [9, 13, 14] to the *relative-saturation* property are also relevant in this connection. Precise definitions of these and other properties referred to in this introduction will be offered in the sequel.

[14]This is done on the basis of the fact that there exists a measurable mapping h from an abstract atomless probability space to the usual Lebesgue unit interval such that its induced distribution is the Lebesgue measure itself; see [16, Lemma 2.1] and the discussion in Sect. 2 below.

[15]This is a consequence of the well-known fact that there exists a continuous onto function from any uncountable compact metric space to $[-1, 1]$; see, for example, Rath–Sun–Yamashige referenced in [25] for this.

come up in regard to the existence of a PSNE for *all* private information games. With Proposition 1 relating to KRS-like games in place, we turn to what we explicitly identify as measure spaces satisfying the *d-property with respect to a measurable, measure-preserving function,* and thereby with respect to a sub-σ-algebra. Such a property of a probability space allows a measurable selection to be chosen from the so-called *d-correspondence,* and one that induces a uniform measure on the range of the correspondence.[16] This property is motivated by a recent consideration in the mathematical literature of correspondences that do not admit measurable selections with pre-specified properties when based on the Lebesgue interval, but do so under an extended Lebesgue interval that goes back to Kakutani in the forties, and one whose σ-algebra is countably-generated.[17] Sect. 3.4 places on the record sufficiency results for the existence of PSNE in KRS-like games, Theorem 1 and its three corollaries based on the d-property. The prominence that we give to measure spaces having the d-property is, to be sure, new to the literature: it undergirds the principal results of this entire work.

Section 4 is in keeping with the expositional thrust of this essay. It presents a leisurely introduction to the construction of the Lebesgue extension based on a 1944 lemma of Kakutani's, and originally introduced in [26]. Since this essay is motivated to the non-expert reader, we also provide an exposition of the construction of the Lebesgue interval based on Carathéodory's theorem. This material of is of course standard.

Section 5 is the dialectical backbone of the paper, and its dynamic turns on two sharp questions, the first of which is the following.

(a) Does the extended probability space, an extension of the Lebesgue interval, resolve the KRS counterexample?

Based on the intuitive discussion of the extension in Sects. 4, 5.1 answers this question in the affirmative. There exists a PSNE in the KRS example *if* the information spaces are modeled on the extension of the Lebesgue interval, rather than on the Lebesgue interval itself. And so this appears to be all that there is to it.[18] Unfortunately, this success is more illusory than real. We show in Sect. 5.2 that the KRS

[16]This correspondence is reproduced in Fig. 2 below, and was referred to in [26] as the *Debreu correspondence* simply as a mnemonic; and as indicated there, Hart-Kohlberg ascribe it to Debreu in an entirely different context and for an entirely different purpose. Our current use of the letter d for this correspondence, and for the d-property of a measure space based on it, is meant to indicate a situation where each type of agent has a dual best-response. However, if the reader wishes, he or she can capitalize d and make a non-obligatory nod in Debreu's direction.

[17]One of these correspondences is precisely the d-correspondence. Another derives from the celebrated example of Lyapunov; see Claims 1–3 in [26, Sect. 1]. We underscore for the general reader the intuitively-obvious fact that the Lebesgue extension is mathematically much simpler than the saturated extension of the Lebesgue interval in [45]. For Lyapunov's theorem, see [23] and their references.

[18]This, by itself is no longer surprising. It is now understood, at least by the *cognoscenti*, that one only needs a σ-algebra that is finer than the Lebesgue σ-algebra in the sense that it contains a set of measure 1/2 and which is independent of the Lebesgue σ-algebra; see [13] written subsequent to the first version of this paper.

example can be modified and resituated on the extended information spaces to yield another troublesome counterexample without a PSNE. This example is of a finite game with information spaces "richer" than those used in the KRS example, but with the payoffs suitably modified and refined to pertain to these spaces. It is this upgrading of the (counter)example that motivates both a KRS-like game and measure spaces satisfying the d-property. In any case, one can now reformulate/repeat the question under discussion, and ask:

(b) Does a further extension of the extended probability space resolve this "new" counterexample?

Perhaps somewhat surprisingly, the answer is again affirmative in that the techniques of [26], and recapitulated in Sect. 3, are up to the task. However, a recursion now suggests itself and is indeed executable in the form of a general result. Even though a finite game Γ_n based on an n-fold extension of the Lebesgue interval has no Nash equilibrium, we can construct an $(n+1)$-fold extension of the information spaces for which it has an equilibrium! And *none* of these constructed games Γ_n can have Nash equilibria in any of the sub-extensions. The point is that all these constructed games are KRS-like games with their information spaces satisfying the d-property.[19] Indeed, this recursive non-existence property culminates in a general theorem; see Proposition 2 in Sect. 5.3 below. The question then is what works? how can this unfortunate recursion be terminated? And it is at this point that our exposition leads to the punchline that we want to express. The dialectic can only be terminated when one relies on the idealized limit of a *saturated* space, or a space satisfying the *relative diffuseness* property. These observations relating to the results of [27], and their extension in [13], constitute the two-paragraphed Sect. 5.4.

The final substantive section of this essay concerns recent work of He–Sun–Sun [12, 14]. In a comprehensive treatment, the authors have posed the question as to "which measure spaces are most suitable for modeling many economic agents?" They propose a class of measure spaces that they refer to as satisfying a condition they term "nowhere equivalent." Thus their work represents the next stage of the ongoing trajectory that we have tried to sketch in this introduction: one that begins with the Lebesgue interval and includes an atomless Loeb space. However, the authors principal focus is on large games and economies, and they do not consider the relevance of their novel concept to finite Bayesian games of private information, though they are undoubtedly aware of how their basic argumentation would extend to this setting. In Sect. 6, we consider how the Lebesgue extension can also be used to resolve a

[19]There is of course a Godelian parallel here. Let T_1 be a suitable theory, which is to say, complete and consistent. Then it admits an undecidable proposition, call it S_1. Let T_2 be T_1 extended by S_1, and denoted $T_2 = \{T_1 + G_1\}$. Observe that although G_1 is trivially deducible in T_2, there is another undecidable in T_2, say S_2 etc. S_n is never decidable in T_{n-1}. In fact there is a countably-infinite series of pairs of *theories and undecidables*! Extensions of this type never work to furnish a general theory. The authors are grateful to Josh Epstein for bringing the relevance of Godel's incompleteness theorem to their attention. Josh also singled out parallels to Galois theory whose pursuit in this paper would have taken us too far afield.

question raised in [14], the role that the dialectic that we have identified here also plays in this setting.

We conclude the paper in Sect. 7 with two further remarks, and with an Appendix that collects the purely technical arguments of this essay.

2 The Model

A *private information game* with independent types consists of a finite set of ℓ players, each of whom (say i) chooses actions from a compact metric space A_i which is not necessarily finite, and has access to (private) information and events, represented by a measurable space $(T_i, \mathcal{T}_i)$, and known only to him, and not necessarily to the other players. This information, or type, is independently drawn among players, moreover, the its distribution forms is a probability measure μ_i on $(T_i, \mathcal{T}_i)$ that is known to all players. We refer to $\{(T_i, \mathcal{T}_i, \mu_i) : i = 1, \ldots, \ell\}$ as the *private information structure* of the game. The private information structure is called *diffused* if for every i, μ_i is an atomless probability measure. We shall follow convention and denote the product $\Pi_{j=1}^{\ell} A_j$ by A, and $\Pi_{j \neq i} A_j$ by A_{-i}.

The payoff function of player i is $u_i : A \times T_i \to \mathbb{R}$, and it depends on the actions chosen by all the players and on his own private information $t_i \in T_i$. We consider the following assumption on the payoff function.[20]

Assumption 1 For each player i,

(i) $u_i(\cdot, t_i)$ is a continuous function on A for every $t_i \in T_i$;
(ii) for each $a \in A$, $u_i(a, \cdot)$ is $\mathcal{T}_i$-measurable on T_i;
(iii) u_i is integrably bounded, that is, there is an integrable function ϕ_i on $(T_i, \mathcal{T}_i, \mu_i)$ such that $|u_i(a, t_i)| \leq \phi_i(t_i)$ holds for each $a \in A$.

We denote a Bayesian game with independent private information by

$$\Gamma = \{(T_i, \mathcal{T}_i, \mu_i), A_i, u_i \ : i = 1, \ldots, \ell\}.$$

A *pure strategy* of a Bayesian game Γ is a $\mathcal{T}_i$-measurable mapping from T_i to A_i. A pure strategy profile $f = (f_1, \ldots, f_\ell)$ of a Bayesian game Γ is a *pure-strategy Nash equilibrium* (PSNE) for the game Γ if for every player i, f_i yields the maximal expected utility when the other players choose f_{-i}.

In the reminder of this section, we turn to the state-of-the-art results on the existence of PSNE in Bayesian games of independent private information that will serve as the backdrop for the results presented in this paper. In terms of background, the original existence RR results on games with finite moves, as in [18, 34, 40], were generalized first to games with countably-infinite moves, and then to those with uncountably infinite ones; see [25] for discussion and basic references. The latter

[20] We work with the simplest model; for extensions, see [7, 8, 19, 21, 39].

generalization invoked an atomless Loeb probability space as the formalization of the space of private information. In [27], the authors show that a saturated private information structure is sufficient for the existence of PSNE in private information games.[21] More interestingly, they also show that this saturation property is also necessary in the sense that if at least two players' private information spaces are modeled by non-saturated spaces, there is a private information game without a PSNE! As such, it closes the circle.[22]

In subsequent work, the sufficiency result has been generalized in an interesting way to which we turn. He–Sun [13] make a distinction between the aspects of information with respect to which a player chooses a particular strategy as opposed to those which lead his or her payoff functions to change. Following [14], they propose the concept of *relative-diffuseness* in Bayesian games. Given a private information structure $\{(T_i, \mathcal{T}_i, \mu_i) : i = 1, \ldots, \ell\}$ with respect to which the players take strategies, let $\mathcal{F}_i$ be the smallest sub-σ-algebra of $\mathcal{T}_i$ with respect to which this player's payoff function is measurable. This private information structure is called **relatively diffused** if $\mathcal{F}_i$ is **setwise coarser** than $\mathcal{T}_i$ in the sense that for every $S \in \mathcal{T}_i$ with positive μ-measure, there exists an $\mathcal{T}_i$-measurable subset $S' \subseteq S$ such that $\mu\left(S'\Delta S''\right) > 0$ for any $S'' \in \mathcal{F}_i^S$ where $S'\Delta S'' = (S'\backslash S'') \cup (S''\backslash S')$. For simplicity, we call $\{(T_i, \mathcal{T}_i, \mu_i), \mathcal{F}_i : i = 1, \ldots, \ell\}$ a *relative private information structure.* This leads to a natural variation of Assumption 1.

Assumption 1$'$ Conditions (ii) and (iii) in Assumption 1 are changed to

(ii)$'$ for each $a \in A$, $u_i(a, \cdot)$ is $\mathcal{F}_i$-measurable on T_i;
(iii)$'$ u_i is integrably bounded, that is, there is an $\mathcal{F}_i$-integrable function ϕ_i on $(T_i, \mathcal{F}_i, \mu_i)$ such that $|u_i(a, t_i)| \leq \phi_i(t_i)$ holds for each $a \in A$.

We now denote a Bayesian game with relatively diffused independent private information by

$$\Gamma = \{((T_i, \mathcal{T}_i, \mu_i), \mathcal{F}_i), A_i, u_i \ : i = 1, \ldots, \ell\}.$$

As before, a pure strategy for player i is still a $\mathcal{T}_i$-measurable function from T_i to her action space A_i. In a phrase, payoffs functions hinge on $\mathcal{F}_i$ and strategies on $\mathcal{T}_i$.

It is then shown in [13, Theorem 1] that there exists a PSNE in a Bayesian game satisfying Assumption 1$'$ if the information on which the players condition their actions is relatively diffused with respect to the information related to the payoffs. This result is a generalization of the sufficiency result in [27, Theorem 1] based on saturated probability spaces. These require that for any nonnegligible subset $S \in \mathcal{T}$, the re-scaled probability space $\left(S, \mathcal{T}^S, \mu^S\right)$ is not essentially countably-generated.[23] Since the σ-algebra generated by any given payoff function satisfying Assumption 1

[21] Since Loeb spaces are saturated, the sufficiency result generalizes previous work; see [27].

[22] It is worthy of emphasis here that, as noted in [27], this necessary and sufficient result was already conjectured in [16]. Indeed, the necessity claim was first made by Keisler–Sun in 2002; see the relevant footnote in their paper.

[23] Here $\mathcal{T}^S$ is the σ-algebra $\left\{S \cap S' : S' \in \mathcal{T}\right\}$ and μ^S is defined on $\mathcal{T}^S$ by $\mu(\cdot) = \mu(\cdot)/\mu(S)$. The reader is referred [27] for details and references.

is always setwise-coarser than the underlying σ-algebra of the saturated space, the relative diffuseness assumption is automatically fulfilled. It is worth underscoring, however, that there is no necessity result in [13].[24]

3 KRS-Like Games Revisited

KRS-like games are studied in [28], these games are constructed based on an example of two-player private information games in [17], and now referred to as the KRS example. It is a two-player, private information game satisfying Assumption 1 with a common action set of uncountable cardinality. Each player $i = 1, 2$, can take actions from $A_i = [-1, 1]$. Let $(T_i, \mathcal{T}_i, \mu_i), i = 1, 2$ be two atomless probability spaces, and let $h_i : T_i \to [0, 1]$ be a $\mathcal{T}_i$-measurable mapping such that the induced distribution over $[0, 1]$ is the Lebesgue measure η.[25] The payoff functions for both players are given as below:

$$u_1^{h_1}(a_1, a_2, t_1) = u_1(a_1, a_2, h_1(t_1)) = -|h_1(t_1) - |a_1|| + [h_1(t_1) - a_1] \cdot z(h_1(t_1), a_2), \tag{1}$$

$$u_2^{h_2}(a_1, a_2, t_2) = u_2(a_1, a_2, h_2(t_2)) = -|h_2(t_2) - |a_2|| - [h_2(t_2) - a_2] \cdot z(h_2(t_2), a_1); \tag{2}$$

where the function $z : [0, 1] \times [-1, 1] \to \mathbb{R}$ is defined as follows. For all $t \in [0, 1/2], b \in [-1, 1]$

$$z(t, b) = \begin{cases} b, & \text{if } 0 \leq b \leq t; \\ t, & \text{if } t < b \leq 1; \\ -z(t, -b), & \text{if } b < 0; \end{cases}$$

and for any $t \in (1/2, 1]$, $z(t, \cdot) = z(1/2, \cdot)$; see Fig. 1 for the graph of $z(t, \cdot)$ for three different values of t.

In particular, $(T_i, \mathcal{T}_i, \mu_i)$ are the usual Lebesgue unit intervals and h_i are the identity maps on $[0, 1]$, this KRS-like game is the original KRS game in [17]. It is also clear that for each player i, and for each $t_i \in T_i$, $u_i(\cdot, \cdot, t_i)$ is a continuous function on $[-1, 1] \times [-1, 1]$. For different $t_i \in T_i$, all $u_i(\cdot, \cdot, t_i)$ constitute an equicontinuous family. Thus, for any $a_1, a_2 \in [-1, 1]$, $u_i(a_1, a_2, \cdot)$ is $\mathcal{T}_i$-measurable function. As a result, KRS-like games satisfy Assumption 1.

---

[24] Even though it constitutes a rather narrow perspective from which to view this paper, one could in principle, see the results reported here as addressing themselves to the problem left open in [13].

[25] There always exists such a function h_i, see [2, Proposition 9.1.11].

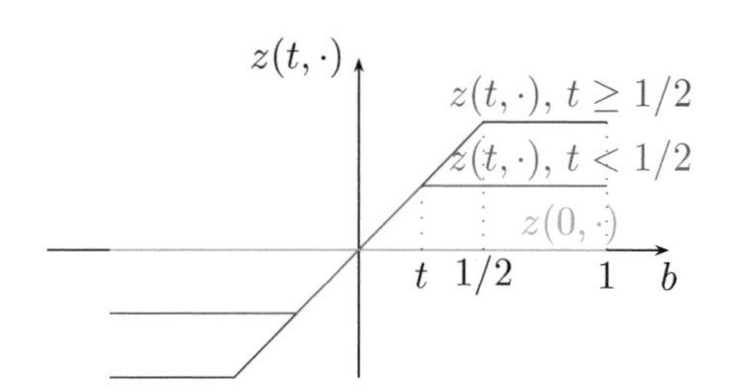

Fig. 1 Graph of $z(t, \cdot)$ for different t

When the private information spaces are given, we denote this game by

$$\Gamma_{h_1,h_2} = \left\{(T_i, \mathcal{T}_i, \mu_i), A_i = [-1, 1], u_i^{h_i} : i = 1, 2\right\}.$$

In this games, for each i, a pure strategy of player i is a $\mathcal{T}_i$-measurable function from T_i to $[-1, 1]$. If (g_1, g_2) be pure-strategy profile, and $\nu_i = \eta g_i^{-1}$ is the induced distribution on the action set $[-1, 1]$.

We are now ready to introduce the equilibria in KRS-like games. First for any $t \in [0, 1]$, any Borel probability measure ν on $[-1, 1]$, let $w(t, \nu)$ be the integral of $z(t, \cdot)$ with respect to ν, i.e.,

$$w(t, \nu) = \int_{-1}^{1} z(t, \cdot)\mathrm{d}\nu. \tag{3}$$

The best-response correspondence in the KRS-like game Γ_{h_1,h_2} is as follows:

$$B_1(t_1; \nu_2) = \begin{cases} -h_1(t_1) \text{ or } h_1(t_1), & \text{if } w(h_1(t_1), \nu_2) = 0; \\ h_1(t_1), & \text{if } w(h_1(t_1), \nu_2) > 0; \\ -h_1(t_1), & \text{if } w(h_1(t_1), \nu_2) < 0. \end{cases}$$

$$B_2(t_2; \nu_1) = \begin{cases} -h_2(t_2) \text{ or } h_2(t_2), & \text{if } w(h_2(t_2), \nu_1) = 0; \\ -h_2(t_2), & \text{if } w(h_2(t_2), \nu_1) > 0; \\ h_2(t_2), & \text{if } w(h_2(t_2), \nu_1) < 0. \end{cases}$$

Moreover, suppose that there exists a PSNE (g_1^*, g_2^*) in the game Γ_{h_1,h_2}, where g_i^* is a $\mathcal{T}_i$-measurable function from $(T_i, \mathcal{T}_i, \mu_i)$ to $[-1, 1]$. Let ν_i^* be the induced distribution of g_i^*, i.e., $\nu_i^* = \mu_i \circ (g_i^*)^{-1}$. The equilibrium action distribution of each player's strategy also satisfies the following statement.

Proposition 1 (Khan-Zhang [28]) *Suppose that ν_1^*, ν_2^* are the induced action distributions of a PSNE of the game Γ_{h_1,h_2}, then for $i = 1, 2$, $w(h_i(t_i), \nu_i^*) = 0$ for μ_i-almost all $t_i \in T_i$, and $\nu_i^*([0, s]) = \nu_i^*([-s, 0]) = s/2$ for any $s \in [0, 1/2]$.*

Khan and Zhang also find that PSNE in KRS-like games is intimately related to the following d-property concept of probability spaces.

Definition 1 (i) Given an atomless probability space $(T, \mathcal{F}, \mu)$ and a $\mathcal{F}$-measurable measure-preserving map h from T to the Lebesgue interval $([0, 1], \mathcal{L}, \eta)$,[26] $(T, \mathcal{F}, \mu)$ is said to have the **relative d-property** with respect to h if there is a $\mathcal{F}$-measurable map g from T to $[-1, 1]$ such that $g(t) \in \{h(t), -h(t)\}$ and g induces the uniform distribution on $[-1, 1]$. (ii) Given an atomless probability space $(T, \mathcal{G}, \mu)$ where $\mathcal{G}$ is a sub-σ-algebra of $\mathcal{F}$, $(T, \mathcal{F}, \mu)$ is said to have the **relative d-property** with respect to $\mathcal{G}$ if for all $\mathcal{G}$-measurable measure-preserving map h from T to the Lebesgue interval $([0, 1], \mathcal{L}, \eta)$, $(T, \mathcal{F}, \mu)$ has the relative d-property with respect to h.

It is clear that the usual Lebesgue unit interval $([0, 1], \mathcal{L}, \eta)$ does not have relative d-property with respect to the identity map on the interval, which is obviously a measure-preserving map. The following is straightforward from Proposition 1.

Corollary 1 *For $i = 1, 2$, if an atomless probability space $(T_i, \mathcal{T}_i, \mu_i)$ has relative d-property with respect to a measure preserving map h_i from T_i to the Lebesgue interval, then there exists a pure-strategy Nash equilibrium in the KRS-like game Γ_{h_1,h_2}.*

Here is a sufficient condition for the relative d-property.

Lemma 1 *Given $(T, \mathcal{G}, \mu)$, $\mathcal{G}$ as in Definition 1. If there is a $\mathcal{F}$-measurable subset with μ-measure (1/2), and it is independent with $\mathcal{G}$, then $(T, \mathcal{F}, \mu)$ has the relative d-property with respect to $\mathcal{G}$.*

Moreover, the relative d-property of a probability space also furnishes a necessity condition for modeling the private information spaces such that KRS-likes games all have PSNE.

Theorem 1 (Khan-Zhang [28]) *Given a diffused private information structure $\{(T_i, \mathcal{T}_i, \mu_i) : i = 1, 2\}$, if for some i, $(T_i, \mathcal{T}_i, \mu_i)$ does not have relative d-property with respect to a measure preserving map h_i from T_i to the Lebesgue unit interval, then there exists a KRS-like game possessing no pure-strategy equilibrium.*

4 A Lebesgue Extension à la Kakutani

In this section, we attempt to lay out for the general reader the basic intuitions underlying the construction of the Lebesgue extension rather than simply using it as a black-box that furnishes a pure-strategy equilibrium in a class of games that do not possess such an equilibrium. To put the point another way, the principles underlying the extension go beyond the technical to the substantive considerations.

To be sure, the extension of the Lebesgue measure has by necessity to build on the construction of the Lebesgue measure itself, and we begin this section by recalling

[26] A map $h : (T, \mathcal{F}, \mu) \to ([0, 1], \mathcal{L}, \eta)$ is called measure-preserving if h is $\mathcal{F} - \mathcal{L}$-measurable and the induced distribution of h is the Lebesgue measure on the unit interval.

the basic (standard) principles underlying this construction.[27] Towards this end, we begin by recalling the notion of an *outer measure* θ on the power set $\mathcal{P}(X)$ of a space X. This is simply a non-negative function that gives zero value to the empty set, is *monotonic* and *countably subadditive.* This is to say

$$\theta(\emptyset) = 0, \;\; A \subseteq B \Longrightarrow \theta(A) \leq \theta(B) \text{ and } \theta(\cup_{n \in \mathbb{N}} E_n) \leq \sum_{n \in \mathbb{N}} \theta(E_n).$$

If the outer measure of X is unity, then it is a pre-probability, and what one needs to get a *bona fide* probability is to strengthen countable subadditivity to countably additivity. The point is that on restricting an outer measure to a specific class of subsets of X, this can be done and it turns into a measure. It is important to understand this restricted class of sets, and we turn to it.

It is clear that any set A can be disjointly decomposed through another set B by viewing it as the intersection of it with E and the set of its points that do not belong to E. In symbols,

$$A = (A \cap E) \cup (A/E) \text{ where } A, E \in \mathcal{P}(X).$$

We can refer to E as a *decomposing* set, and the sets $A \cap B$ and A/E as its decompositions with respect to it. This much is entirely trivial.[28] Now focus on a set E that decomposes any subset of $\mathcal{P}(X)$ in a way that the outer measure of the set and the sum of the outer measures of the its decompositions with respect to E are identical. This is to ask for a focus on

$$\sum = \{E \in \mathcal{P}(X) : \theta(A) = \theta(A \cap E) + \theta(\cup(A/E)) \text{ for all } A \in \mathcal{P}(X).$$

Now what is not trivial is that $\sum$ is a σ-algebra, which is to say in the language of probability theory, a *bona fide* event space: closed under complementation and countable unions. And more to the point, the outer measure θ restricted to this class is a measure which is to say countably additive for a disjoint sequence of events. Again, in the restricted language of probability theory, a pre-probability has been rendered by restriction to a probability, a result that goes by the name of Carathéodory.[29]

So far, in the consideration of an abstract set X, we have had nothing to say as regards a Lebesgue measure. Indeed, we have simply specified a methodology by which a given outer measure on a power set can be turned into a measure on a specific distinguished σ-algebra chosen from that power set. We now specialize X to $\mathbb{R}$, and rather than assume an outer measure, construct it. For any two real numbers

[27] Several excellent treatments of this standard material are available, but we hope that the following two paragraphs will not only set the stage for what is to follow but will speak to the general reader; for details, we recommend [6, 37].

[28] But see Nillson's singling this equality out in [37, Eq. 5.6, p. 304]. Khan would like to take this opportunity to thank Metin Uyanik for discussion concerning this "Carathéodory equation."

[29] See the epigraph, and the discussion in [37, Sect. 5.4].

a and b, consider as a building block the half-open interval $\{x \in \mathbb{R} : a \leq x < b\}$, and associate with it the number $b - a$ when $b \geq a$, and zero otherwise. Refer to this association as a function ℓ on half-open intervals on $\mathbb{R}$. This has the intuitive property[30] that the number associated with any half-open interval I, $\ell(I)$, is not greater than the sum of the numbers associated with any countable cover of half-open intervals I_j, $j \in \mathbb{N}$, which is to say, $\sum_{j\in\mathbb{N}} \ell(I_j)$. Two points need to be noticed: the statement pertains only to half-open intervals and to any countable cover of them, much less the most parsimonious one. As a consequence, λ is not yet an outer measure on $\mathcal{P}(\mathbb{R})$.

It is now a straightforward matter to use the function λ defined on the basic building blocks to construct a function θ defined on $\mathcal{P}(\mathbb{R})$ by limiting ourselves to the most parsimonious cover of an arbitrary subset A in $\mathbb{R}$. The symbolism is transparent:

$$\theta(A) = \inf_{\{I_j\}_{j\in\mathbb{N}}} \left\{ \sum_{j\in\mathbb{N}} \ell(I_j) : A \subseteq \bigcup_{j\in\mathbb{N}} I_j \right\},$$

though it bears emphasis that the infimum is taken over all countable covers of A. The fact that θ gives the zero value to the empty set, and that it is monotonic is a triviality; in order to show that it is countably subadditive is a routine computation drawing what it means to have an infimum.[31] But now we can appeal to Carathéodory's procedure to obtain a distinguished σ-algebra in $\mathcal{P}(X)$, the (Lebesgue σ-algebra) and a measure (the Lebesgue measure) on it. This measure space furnishes the Lebesgue unit interval when it is restricted to the unit interval; and it is the extension of this space $(L = [0, 1], \mathcal{L}, \eta)$ that is at issue.

Thus, consider the Lebesgue unit interval, $L = ([0, 1], \mathcal{L}, \eta)$ as the primitive object to which we seek an extension. Even a cursory perusal of the argumentation underlying the construction of the extended Lebesgue interval shows its dependence on a 1944 Lemma of Kakutani, [15]. To facilitate the intuition behind the procedures of this extension, think in terms of an allegory of a "treasure hunt" in which one is to find bills of denomination ranging from zero to one, $K = [0, 1]$, buried in locations proxied by numbers between zero to one, $L = [0, 1]$. The set of locations is already furnished with a Lebesgue measure-theoretic structure: this is to say that we have assumed an ability to measure the length of any interval between two locations (ℓ, ℓ'). Let us now also explicitly assume a Lebesgue measure-theoretic structure $([0, 1], \mathcal{K}, \kappa)$ on the set of denominations K, and seek to estimate a measure of the size of the 'treasure"– the total of the amount given by the bills of denomination between (k, k') and buried in the strip of land between (l, l'). To repeat, we aim here for an exposition that gives the basic intuition behind the construction, and refer any reader interested in the details of the rigorous argumentation to [26].

[30] Even though the property is intuitive, relying as it does on the notion of a *length* of an interval and what it means to have cover, it nevertheless requires a proof. Henceforth, by *cover* we shall mean a cover of half-open intervals.

[31] See the notes and comments in Fremlin ([6]; Sect. 113); also see [3, 37].

Kakutani's Lemma: *There exists a partition of uncountable cardinality of* $L = [0, 1]$, *denoted by* $\{C_k : k \in K = [0, 1]\}$, *such that the Lebesgue outer-measure*[32] *of* C_k *is one for all* $k \in K = [0, 1]$.

Now Kakutani's lemma furnishes the rudiments from which a "treasure map" C in the space of all the locations and the denominations can be constructed. The lemma furnishes a *partition* of the unit interval indexed by each denomination. Heuristically, every location is assigned a unique amount of wealth, and the location of bills with a given denomination level k is rather dispersed among the set of all locations. Symbolically, we are furnished with $\{C_k \subseteq [0, 1] : k \in K\}$ such that $\cup_{k\in[0,1]} C_k = [0, 1]$ and $C_i \neq C_j,\ i \neq j$. However, the point is that the length of these C_k cannot be determined: none of them are in general Lebesgue measurable, but only Lebesgue outer-measurable, each with unit *outer-measure*. This is to say that the "smallest" Lebesgue measurable set containing a given C_k has a unit Lebesgue measure. We can now take this partition and "unfold" it as the "treasure-map" C where

$$C = \cup_{k\in[0,1]} C_k \times \{k\} \subseteq L \times K.$$

The point is that this set C is also only outer-measurable with unit (square) Lebesgue measure $\eta \otimes \kappa$. The "treasure map" is not accurate enough!

To overcome this deficiency, consider the σ-algebra generated by C and the sets in $\mathcal{L} \otimes \mathcal{K}$, and extend the square Lebesgue measure $\eta \otimes \kappa$ to this extended σ-algebra $\mathcal{T}$. Denote this extended measure by γ, and note that we have a measure-theoretic structure, $(C, \mathcal{T}, \gamma)$, on C such that the σ-algebra $\mathcal{T}$ is the restriction of the Lebesgue product σ-algebra $\mathcal{L} \otimes \mathcal{K}$ on C. This takes us to the second foothold of the extension procedure. It is simply that the size of any set of the form $((l, l') \times (k, k')) \cap C$ with respect to γ inherits the size of the rectangle $(l, l') \times (k, k')$ in the square. This is to say that for all $0 \leq l < l' \leq 1$ and $0 \leq k < k' \leq 1$,

$$\gamma\left[((l, l') \times (k, k')) \cap C\right] = (l' - l)(k' - k).$$

Finally, we project the unit square to the unit interval. This is to say that we consider the projection p from C to L, and observe it to be a one-to-one measurable mapping. Hence it induces a probability structure on [0, 1] by projecting the probability structure on C.

Denote the new probability structure on [0, 1] by $([0, 1], \mathcal{I}, \lambda)$, and this is the extension of the Lebesgue unit interval that we seek. It is now worthwhile to summarize the procedure. Each type has a double identity: an explicit identity or trait (say, e.g., the location) indexed by elements of L and another implicit identity or trait (say, e.g., the wealth level) indexed by elements of K, and the two traits co-exist in single-dimensional set I. The point of consequence is that these two traits are governed by two independent σ-algebras, and the extended Lebesgue

[32]Given a measure space $(T, \mathcal{T}, \mu)$, the associated outer measure, denoted by μ^*, is defined as follows: for any subset $E \subseteq T$, $\mu^*(E) = \inf\{\Sigma_n \mu(E_n) : E_n \in \mathcal{T}, E \subseteq \cup_n E_n\}$, it bears emphasis that the infimum is taken over all countable covers of E.

measure is atomless on both. Next, we turn to this. For all $0 \leq k < k' \leq 1$, let $D_{kk'} = \cup_{k<k''<k'} C_{k''}$, which is the set of all implicit traits lying between k and k'. Notice that $p^{-1}(D_{kk'}) = \big([0, 1] \times [k, k')\big) \cap C$, and by virtue of the way that the extended σ-algebra $\mathcal{I}$ was obtained on I, $D_{kk'} \in \mathcal{I}$. Furthermore, by virtue of the way that the extended Lebesgue measure was obtained on $\mathcal{I}$, we have

$$\lambda(D_{kk'}) = \gamma\left[p^{-1}(D_{kk'})\right] = \gamma\left[\big([0, 1] \times (k, k')\big) \cap C\right] = k' - k.$$

That is, the probability of a type whose implicit trait lies between k and k' is exactly $k' - k$. That is, the wealth level, viewed as a random variable on the extended Lebesgue interval, is a measurable measure-preserving map to the Lebesgue interval.

Next, we claim that the two random variables, the wealth level and the location, are independent. Fix $0 \leq k < k' \leq 1$ and $0 \leq l < l' \leq 1$, consider the probability of types where the wealth lies between k and k' and the location lies between l and l'. Independence of the two random variables simply means that the probability of the types that lie in the intersection of the two sets is the product of the probability that the type lies in each of the sets. But this clear on account of the fact that $p^{-1}\left(D_{kk'} \cap (l, l')\right) = \big((l, l') \times (k, k')\big) \cap C$, and thus

$$\lambda\left(D_{kk'} \cap (l, l')\right) = \gamma\left[p^{-1}\left(D_{kk'} \cap (l, l')\right)\right] = \gamma\left[\big((l, l') \times (k, k')\big) \cap C\right] = (l' - l)(k' - k). \tag{4}$$

We thus completes the proof of the independence between the wealth level and the location.

In summary, the extension proceeds in the following steps: (i) the Kakutani partition of the Lebesgue unit interval, (ii) the lifting of this partition to a set C in the Lebesgue square, (iii) the extension of the square Lebesgue measure-theoretic structure to include C, (iv) restriction of this structure to C, and finally, (v) a projection of this restricted structure to the given Lebesgue interval.[33] The point is that one can now estimate the size of many more sets by λ than we could before.

Once this extension is understood, only a little more is involved in understanding that a sequence of Lebesgue extensions $\{([0, 1], \mathcal{I}_n, \lambda_n) : n = 0, 1, \ldots\}$ can be constructed in which the first countably-generated Lebesgue extension $([0, 1], \mathcal{I}, \lambda)$ is denoted by $([0, 1], \mathcal{I}_0, \lambda_0)$, and for any $n \in \mathbb{N}$, $([0, 1], \mathcal{I}_n, \lambda_n)$ is an extension of $([0, 1], \mathcal{I}_{n-1}, \lambda_{n-1})$, where the former is obtained from the latter in precisely the way that $([0, 1], \mathcal{I}, \lambda)$ is obtained from the Lebesgue interval.[34] We can now record the following properties of these extensions.

[33]The details of each of these steps are spelt out in [26]. It is a good exercise for the interested reader to work out for herself the routine arithmetic behind each of these steps. She should note, in particular, that the proof of the claim that the outer-measure of C is unity (straightforwardly) invokes Fubini's theorem.

[34]As in Footnote 32, we send the reader interested in the details to [26]; and in this particular context, to Sects. 5.2.2 and 5.2.3 in that paper.

Lemma 2 *(i) For each $n \in \mathbb{N}$, $\mathcal{I}_{n-1}$ is setwise coarser than $\mathcal{I}_n$. (ii) For every $n \in \mathbb{N}$, there exists an $\mathcal{I}_n$-measurable measure-preserving map h_n such that for any $E \in \mathcal{I}_n$, there exists a Lebesgue measurable subset $S \subseteq [0, 1]$ such that $\lambda_n[E \Delta h_n^{-1}(S)] = 0$ where Δ is the symmetric difference operator in $\mathcal{I}_n$. (iii) The n-fold Lebesgue extension does not have d-property with respect to h_n, for all $n \in \mathbb{N}$. In particular, when restricted to $h_n^{-1}([0, 1/2])$, there is no selection of the corresponding d-correspondence of h_n such that the induced distribution is uniform on $[-1/2, 1/2]$.*

Remark 1 The m-th fold Lebesgue extension has d-property with respect to $\mathcal{I}_n$. We also note that Lemma 2(iii) allow us to assert that no matter how large a natural number n is, the n-fold Lebesgue extension extension is not a saturated space. The point in part (ii) is that $([0, 1], \mathcal{I}_n, \lambda_n)$ does not have the relative d-property with respect to the measure-preserving map h_n.

5 KRS-Like Games Based on Lebesgue Extensions

In this section, we turn to the KRS example itself, and ask whether one can obtain a PSNE in the game Γ_0 by jettisoning the usual Lebesgue unit interval, and turning not to a saturated or super-atomless probability space,[35] but to an atomless probability structure whose measure-theoretic is rich enough *only* to the point that is required to show the existence of a PSNE in the specific game Γ_0. This is to ask for a measure-theoretic structure that is oriented towards resolving and subduing the canonical counterexamples. We develop the answer to this question in Sect. 5.1 by using the countably-generated extension of the usual Lebesgue unit interval offered by the authors in [26]. However, in Sect. 5.2, we show that this "more sophisticated and enriched" atomless probability space generates its own example of finite-player games without a PSNE. As already stated informally in the introduction, this counter-example on the extended information space can in its turn be resolved by a *further* enrichment of the (extended) σ-algebra. In Sect. 5.3, we conclude with a general result formalizing this dialectic. However, prior to all this, we review for the reader the highlights of the construction of the extended Lebesgue interval.[36]

5.1 The KRS Example Resolved

We now turn to the non-existence of a PSNE in the game Γ, and ask whether the use of extended Lebesgue interval as the space of private information resolves the problem.

---

[35] See [4, 16] for definition of these terms.

[36] It may be worth pointing out that this is the first substantive application, and an exposition, of this extended Lebesgue interval in the economics literature.

The affirmative answer to this question can now be routinely outlined. Consider the game,

$$\tilde{\Gamma}_0 = \{(T_i, \mathcal{T}_i, \mu_i) = ([0, 1], \mathcal{I}, \lambda), A_i = [-1, 1], u_i : i = 1, 2\},$$

and note that $\tilde{\Gamma}_0$ is identical to Γ_0 except for the fact that each player's private information space is replaced by the (countably-generated) Lebesgue extension. We have subdued the counterexample and resolved its negativity by this "tilde" operation involving a move from Γ_0 to $\tilde{\Gamma}_0$. This move is important for the argument that is being developed here. We can now present

Claim 1 *There exists a PSNE in the game* $\tilde{\Gamma}_0$.

This result is an easy consequence of Proposition 1 and Lemma 2(i).

However, a natural question arises as to whether a general theorem can be developed for Bayesian games with compact metric actions sets based on information spaces modeled by the extended Lebesgue intervals. as we shall see in the next subsection, the answer is unfortunately resoundingly negative.

5.2 Yet Another Counterexample

In order to develop the counterexample, we return to Lemma 2(ii), and work with the measurable the measure-preserving map h_0 from $([0, 1], \mathcal{I}, \lambda)$ to the usual Lebesgue interval guaranteed therein. Use this map h_0 to formulate the following KRS-like game, Γ_{h_0,h_0}.

$$\Gamma_{h_0,h_0} = \{(T_i, \mathcal{T}_i, \mu_i) = ([0, 1], \mathcal{I}, \lambda), A_i = [-1, 1], u_i^{h_0} : i = 1, 2.\}$$

We can now appeal to Lemma 2(iii) to assert that

Claim 2 *There does not exist a PSNE in the game* Γ_{h_0,h_0}.

But now one is on a roll. One can ask whether the non-existence of a PSNE in the KRS-like game Γ_{h_0,h_0} can be is resolved in precisely the same way that the non-existence issue for the KRS game Γ_0 was resolved. This is to check whether a further extension of the extended Lebesgue interval would subdue the new example. This can be done by a consideration of the following game,

$$\tilde{\Gamma}_{h_0,h_0} = \{(T_i, \mathcal{T}_i, \mu_i) = ([0, 1], \mathcal{I}_1, \lambda_1), A_i = [-1, 1], u_i^{h_0} : i = 1, 2\},$$

in which each player's private information space is "upgraded" from $([0, 1], \mathcal{I}, \lambda)$ to $([0, 1], \mathcal{I}_1, \lambda_1)$. To use the earlier vernacular, the resolution hinges on a further "tilde" operation involving a move from Γ_{h_0,h_0} to $\tilde{\Gamma}_{h_0,h_0}$. We can now again record the following easy consequence of Proposition 1 and Lemma 2(i).

Claim 1′ *There does exist a PSNE in* $\tilde{\Gamma}_{h_0,h_0}$.

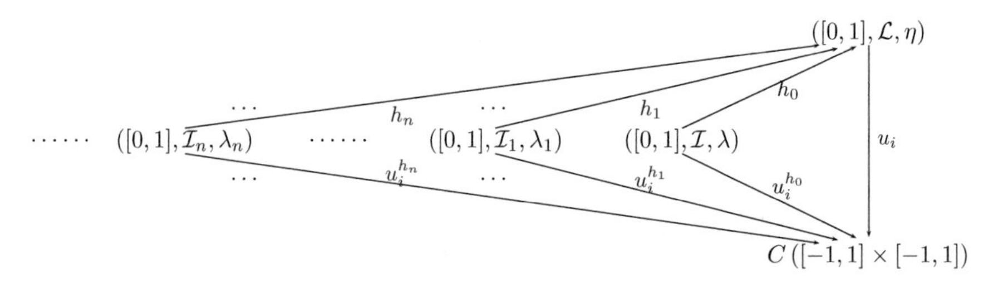

Fig. 2 Lebesgue extensions and KRS-like games

5.3 A General Negative Result

The interesting question is whether there is a general recursion theorem here. We develop such a result in this subsection. The point is that the argumentation in Sect. 5.2 can be continued inductively *ad infinitum*. First, a sequence of countably-generated probability spaces $\{([0, 1], \mathcal{I}_n, \lambda_n) : n = 0, 1, \ldots\}$ can be constructed, where the first countably-generated Lebesgue extension $([0, 1], \mathcal{I}, \lambda)$ is denoted by $([0, 1], \mathcal{I}_0, \lambda_0)$, and for any $n \in \mathbb{N}$, $([0, 1], \mathcal{I}_n, \lambda_n)$ is a countably-generated extension of $([0, 1], \mathcal{I}_{n-1}, \lambda_{n-1})$. Second, if each player's private information space is modeled by $([0, 1], \mathcal{I}_{n-1}, \lambda_{n-1})$, there exists a KRS-like game $\Gamma_{h_{n-1},h_{n-1}}$ without any PSNE. Such that there does not exist a PSNE. Third, as far as this KRS-like game is concerned, the "tilde" operation whereby each player's private information space is modeled by the countably-generated space $([0, 1], \mathcal{I}_n, \lambda_n)$, again guarantees a PSNE. This argumentation can be succinctly illustrated and summarized by Fig. 2, where $C\,([-1, 1] \times [-1, 1])$ means the space of all continuous functions on $[-1, 1] \times [-1, 1]$.

In terms of a formal statement, we can offer:

Proposition 2 *For each $n \in \mathbb{N}$, there does not exist a PSNE in the KRS-like game Γ_{h_n,h_n} but there does exist one in the game $\tilde{\Gamma}_{h_n,h_n}$, where the private information space for each player in $\tilde{\Gamma}_{h_n,h_n}$ is upgraded from the n-fold extension to the $(n+1)$-fold extension.*

Proposition 2 embraces both a positive and a negative result, and in conclusion, it is worthy of note that Claims 1 and 2 above follow as its special cases.

5.4 A Discussion of the Results

The positive result in Proposition 2 can be viewed as an illustration of Theorem 1 of [13] since the relative diffuseness assumption in such KRS-like games are satisfied. However, as far as the KRS example and the KRS-like games Γ_{h_n,h_n} are concerned, it follows from Lemma 1 that a rather simpler and more modest extension of the underlying private information space suffices: all one has to do is to include a subset with measure one-half and one that is independent of the underlying σ-algebra. It serves as the "right" model of the private information space.

The negative result in Proposition 2, the non-existence of PSNE in the KRS-like games Γ_{h_n,h_n}, can be viewed as a special case of the necessity result, Theorem 2, in [27]. There it states that if two players' private information spaces are both modeled by non-saturated probability spaces, then there exists a counterexample of a private information game without any PSNE. In the KRS-like game Γ_{h_n,h_n}, the underlying private information spaces are both n-fold Lebesgue extensions, and thereby essentially countably generated spaces, and automatically not saturated spaces.[37] However, the non-existence argument here is different from the one in the proof of Theorem 2 in [27]: here it is a rather straightforward consequence of Proposition 1 and Lemma 2(iii). It is in this regard that the approach used in this paper is self-consistent, as far as the construction of the counterexamples Γ_{h_n,h_n} are concerned.

6 A Condition of He–Sun–Sun

In work circulated in 2013, He et al. have proposed a far-reaching generalization of the saturation property in the form of condition they of *nowhere equivalence* of two σ-algebras of a probability space. They have relied on this condition to present a comprehensive theory of economies and games with a continuum of agents, and of the three basic mathematical operations that undergird it: integration, distribution and conditional expectation. This work is important enough this expository paper would not be complete in its scope without making some reference to this work. In this section we relate the Lebesgue extension and the ideas presented above to this important work.

He–Sun–Sun [14] motivate their condition, and their results based on it, by a series of examples of games and economies which show pathological features as far as the existence, closed graph and "determinateness" properties of the equilibria are concerned. Here we consider Example 3 of [14], henceforth the HSS example. In this example, there are two large games, both have Lebesgue interval as agent space, the common action space is $[-1, 1]$. Moreover, in both games, each player's payoff only depends on her own action, not anybody else's. Namely, for all agent $i \in [0, 1]$, $a \in [-1, 1]$, and any action distribution ν on $[-1, 1]$,

$$G_1(i, a, \nu) = -(a+i)^2(a-i)^2, \text{ and } G_2(i, a, \nu) = \begin{cases} G_1(2i, a, \nu), & \text{if } i \in [0, 1/2), \\ G_1(2i-1, a, \nu), & \text{if } i \in [1/2, 1]. \end{cases}$$

[37] More precisely, in the KRS-like game Γ_{h_n,h_n}, the corresponding s_1, s_2 in the proof of [27, Theorem 2] are both one.

Note that in G_1, player i's best strategy, no matter what the strategy of all others, is always either i or $-i$, while in G_2, the best strategy for Mr i is either $2i$ or $-2i$, for i less than one half, and either $2i - 1$ or $1 - 2i$ for i great than one half. As a result, in G_1, a PSNE will be a Lebesgue measurable map from $[0, 1]$ to $[-1, 1]$ such that the value at i is either i or $-i$, or a Lebesgue-measurable selection of the correspondence $\Phi : [0, 1] \twoheadrightarrow [-1, 1]$ with $\Phi(i) = \{i, -i\}$. Similarly, a PSNE in G_2 is a Lebesgue-measurable selection of the correspondence $\Psi : [0, 1] \twoheadrightarrow [-1, 1]$ with

$$\Psi(i) = \begin{cases} \{2i, -2i\}, & \text{if } i \in [0, 1/2), \\ \{2i - 1, 1 - 2i\}, & \text{if } i \in [1/2, 1]. \end{cases}$$

These two games G_1 and G_2 induce the same distribution on the space of payoff functions, namely space of all continuous functions on the product space of $[-1, 1]$ and all Borel probability measure on $[-1, 1]$. However, the set of action distribution of all PSNE in G_1, denoted by $\mathcal{D}(G_1)$ is not the same as that in G_2, denoted by $\mathcal{D}(G_2)$; see Claim 3 in [14]. Namely, $\mathcal{D}(G_1)$ is the set of induced distribution by all Lebesgue-measurable selection of the correspondence Φ, and $\mathcal{D}(G_2)$ the set of induced distribution by all Lebesgue-measurable selection of the correspondence Ψ. More precisely, the uniform distribution on $[-1, 1]$ is an element of $\mathcal{D}(G_2)$ but not of $\mathcal{D}(G_1)$; see Fig. 3 below.

However, if the agent spaces in these two games are both modeled by the Lebesgue extension as in Sect. 4, and if $\mathcal{D}'(G_1)$ is the set of induced distribution of all $\mathcal{L}^e$-measurable selections of Φ and $\mathcal{D}'(G_2)$ the set of induced distributions by all $\mathcal{L}^e$-measurable selections of Ψ, we obtain the following result.

Proposition 3 *If in this example, both agent spaces are modeled by the Lebesgue extension, then* $\mathcal{D}'(G_2) = \mathcal{D}'(G_1)$.

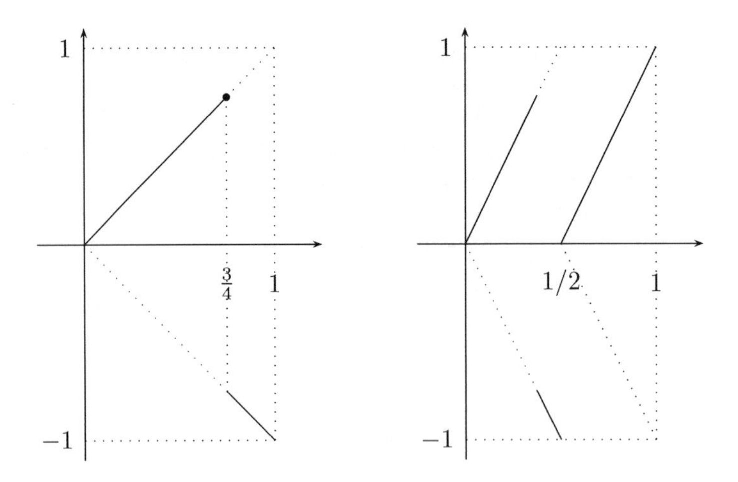

Fig. 3 One selection of Φ and one of Ψ

The proof is postponed to the appendix.

To summarize, the problem raised in Example 3 of [14] automatically disappears when modeling agent space by the Lebesgue extension. This result is not surprising because the extended sigma algebra in the Lebesgue extension satisfies the conditions in Theorem 3 of [14]. The following is a concept proposed in [14].

Definition 2 Given an atomless probability space, $(T, \mathcal{F}, \mu)$, and a sub-σ-algebra $\mathcal{G}$ of $\mathcal{F}$. $\mathcal{F}$ is said to be **nowhere equivalent** to $\mathcal{G}$ if for every $D \in \mathcal{F}$ with $\mu(D) > 0$, there exists a $\mathcal{F}$-measurable subset D_0 of D such that $\mu(D_0 \triangle D_1) > 0$ for any $D_1 \in \mathcal{G}^D$.

In the current context, we only need to claim that the extended Lebesgue σ-algebra in Sect. 5 is nowhere equivalent to the original Lebesgue σ-algebra. The proof of this claim is provided in the Appendix.

We conclude this subsection with two observations. First, we note that the HSS example, and the HSS theorems based on it, concern large games with a continuum of players; and that the reader can generalize the necessary and sufficient results presented in [27, 28] to a finite-player Bayesian games, focused on in the essay, where the analog of the HSS condition is expressed for the spaces of private information. Second, we leave it to the reader to think out for herself a dialectical argumentation underlying the HSS example of the kind that we have considered in his essay for the KRS example.

7 Concluding Remarks

In [27], the authors show that if each player's private information space is modeled by a saturated probability space, then every private information game has a PSNE. As to whether such a saturated private information structure is a "minimal" one for the existence of PSNE in such games, a complete answer is provided in [28] that if every KRS-like game has a PSNE, the underlying information space for each player *must* be saturated. With these two results in hand, under a given private information structure, the hypothesis that all KRS-like games have PSNE implies that all private information games also have PSNE! In other words, KRS-like games are precisely the "trouble-makers" we need to consider and rule-out to establish a general theory on the existence of PSNE for a given private information structure. It is in this sense that we say that KRS-like games serve as a diagnostic tool for the existence of PSNE in private information games.

The dialectic arguments using Lebesgue extensions provide some further elaboration and elucidation on the above "minimal" requirement on the private information structure to guarantee the existence of PSNE. Note that in a saturated probability space, it has a "rich" sigma algebra such that when restricted to any non-negligible subset, the sub-sigma algebra under the restricted measure cannot be essentially countably generated, i.e., there is no hope to find a countable number of sets in the restricted sub-algebra to generate the restricted sub-algebra itself. However, in the n-fold extensions of the Lebesgue interval considered in this essay, the underlying sigma-algebra, no matter how large n is, is essentially countably generated; and, as a result, each n-fold Lebesgue extension cannot be saturated. This is why there is such a KRS-like game without a PSNE as claimed in Proposition 2, a result also implied by the necessity result in [28].

It is worth pointing out another distinction between the Lebesgue extensions considered in this paper and a saturated space. As is made clear in the construction of the Lebesgue extension in Sect. 5, the 1-fold Lebesgue extension, the original Lebesgue σ-algebra is enlarged in a way such that it can accommodate at most two independent random variables, each of which is a measurable measure-preserving from the Lebesgue extension to the Lebesgue interval. Similarly, in the n-fold Lebesgue extension, the underlying σ-algebra can at most accommodate n independent random variables, each of which is a measurable measure-preserving from the n-fold Lebesgue extension to the Lebesgue interval. In comparison, in a saturated probability space, the underlying σ-algebra can accommodate at least a countable number of such pairwise-independent random variables.

We conclude this discussion by an observation that looks at the dialectic of these results from another, and more critical, point of view. The necessity result ensures that for this extended probability space, there will *always* exist a large game without pure-strategy Nash equilibria, but this game may not have any substantive interest. It may be an artifice, a purely technical construction testifying to a mathematical necessity, but with no counterpart in terms of concrete "real-life" applications. Thus, one could legitimately hold the view that as far as the substantive applications are concerned, there is little need for a result that proceeds beyond the modest extension articulated in [26] all the way to a saturated space, or to the spaces satisfying the HSS condition. This is a point of view explicitly articulated in [26] in the context of large non-anonymous games, and further discussion and exploration of whether this is, or is not, only cold comfort for finite games with private information, must be left for future work that turns to concrete applications. For these, see [34, 43] and the references therein, especially to the work of Athey and McAdams. Recall also that the introduction of [34] opens with William Vickrey's auction paper of 1961. The reader should keep this cautionary skepticism in mind now that she has worked through the dialectical argumentation.

8 Technicalities of the Proofs

Proof of Lemma 2. We first prove Part (i). Note that for any natural number n, the n-fold Lebesgue extension is constructed from the $(n-1)$-fold in a similar way as $([0,1], \mathcal{I}, \lambda)$ from the Lebesgue unit interval. As a result, we only need to show that $([0,1], \mathcal{I}, \lambda)$ has the relative d-property for all measure preserving map from the Lebesgue interval to itself.

It follows from Lemma 2 of [26] that there exists a $\mathcal{I}$-measurable subset S such that $\lambda(S) = 1/2$, and both S and S^c is independent with $(0, t)$ for all $t \in [0, 1]$, where S^c is the complement of S. Note that the Lebesgue σ-algebra $\mathcal{L}$ is generated by these subsets $(0, t)$, $t \in [0, 1]$, as a result, both S and S^c are independent with any Lebesgue measurable subset. Given any measure-preserving map h from the Lebesgue interval to itself, let $g : ([0, 1], \mathcal{I}, \lambda) \to [0, 1]$ defined to be $g(t) = h(t)$ for all $t \in S$, and $g(t) = -h(t)$ for all $t \notin S$. It is clear that g is an $\mathcal{I}$-measurable map.

We finally check that g induces the uniform distribution on $[-1, 1]$. For any $s \in [0, 1]$, $\lambda\{t : g(t) \in [-s, 0]\} = \lambda(S^c \cap h^{-1}([0, s])) = \frac{1}{2}\lambda(h^{-1}([0, s])) = \frac{s}{2}$, where the second equation follows from that S^c is independent with the Lebesgue subset $h^{-1}([0, s])$ which is of measure s; similarly, $\lambda\{t : g(t) \in [0, s]\} = \frac{s}{2}$. We thus complete the proof of Part (i).

We next prove Part (ii). The existence of this measure preserving mapping h_n from the n-fold Lebesgue extension to the Lebesgue unit interval is guaranteed by [26, Corollary 1, p. 1093], the key here is that the n-fold Lebesgue extension is an atomless (essentially) countably generated space. ■

Proof of Proposition 2. The existence of pure-strategy Nash equilibria in $\tilde{\Gamma}_{h_n,h_n}$, for all n, follows from Corollary 1 and Part (i) of Lemma 2.

We next prove the non-existence result for the KRS-like game Γ_{h_n,h_n}. By Proposition 1, it suffices to show that there does not exist an $\mathcal{I}_n$-measurable map from $[0, 1]$ to the Lebesgue unit interval such that it takes value either $h_n(t)$ or $-h_n(t)$ for all t, and it induces the uniform distribution on when restricted on $[-1/2, 1/2]$. Suppose not, there is such a mapping g. Let $S = \{t : g(t) \geq 0\}$. It is clear that $S \in \mathcal{I}_n$ and $\lambda_n(S) = \frac{1}{2}$. By the construction of h_n in Lemma 2, there exists an Lebesgue subset E, such that $\lambda_n(S \Delta h_n^{-1}(E)) = 0$. By Part (ii) of Proposition 1, for any $s \in [0, 1/2]$, $\frac{s}{2} = \lambda_n\{t : g \in [0, s]\} = \lambda_n(S \cap h_n^{-1}[0, s]) = \lambda_n(h_n^{-1}(E) \cap h_n^{-1}[0, s]) = \eta(E \cap [0, s])$, where the last equation follows from the measure preserving property. This contradicts the fact that there is no Lebesgue set which is independent with all sets $[0, s]$, $s \in [0, 1/2]$. ■

Proof of the claim in Sect. 6. Let D be an $\mathcal{I}$-measurable subset with $\lambda(D) > 0$. Note that $[0, 1] = D_{01}$, it is clear that there exists two numbers $k, k' \in [0, 1]$, such that $0 < \lambda(D_{kk'} \cap D) < \lambda(D)$. We next fix such a pair of numbers k, k' and construct a required subset $D_0 \in D$ as in Definition 2. Namely, for any subinterval $[l, l'] \subseteq [0, 1]$, $[l, l'] \cap D$ and D_0 differ up to a non-negligible λ-null subset. If this $D_{kk'} \cap D$ and $[l, l'] \cap D$ do not differ up to a null set for all l, l', let $D_0 = D_{kk'} \cap D$. Otherwise, for some l, l', this subset $D_{kk'} \cap D$ and $[l, l'] \cap D$ differ up to a λ-null set. That is, $D_{kk'} \cap [l, l'] \subseteq D$ holds subject to a null set. Let $D_0 = D_{kk''} \cap D = D_{kk''} \cap [l, l']$ where $k'' = \frac{k+k'}{2}$. It is clear that $0 < \lambda(D_0) < \lambda(D)$ and D_0 is a required subset in Definition 2, since $D_{kk''}$ is independent with all subsets $[a, b]$ for all a, b. ■

Proof of Proposition 3. It is clear that $\mathcal{D}'(G_1) \subseteq \mathcal{D}'(G_2)$. What remains is to prove the converse direction. For any distribution μ generated by a $\mathcal{L}^e$-measurable selection of Ψ. Now consider the restricted distribution on $[0, 1]$, still denoted by μ. It is clear that μ is absolutely continuous with respect to the Lebesgue measure on $[0, 1]$, moreover, for any $t \in [0, 1]$, $\mu([0, t])$ is at most of value t. As such, all the conditions of Lemma 3 in [26] are satisfied, therefore, there exists $S_\mu \in \mathcal{L}^e$ such that for any $t \in [0, 1]$, $\mu([0, t]) = \lambda^e(S_\mu \cap [0, t])$. Let f be a $\mathcal{L}^e$-measurable selection of Φ such that $f(i) = i$ if $i \in S_\mu$ and $f(i) = -i$ if $i \notin S_\mu$. It is clear that this f induces the same distribution μ on $[-1, 1]$. ■

References

1. Aumann RJ (2000) Collected papers, vol 1 and 2. MIT Press, Cambridge
2. Bogachev VI (2007) Measure theory, vol II. Springer, Berlin
3. Brucks KM, Bruin H (2004) Topics from one-dimensional dynamics. Cambridge University Press, Cambridge
4. Carmona G, Podczeck K (2009) On the existence of pure-strategy equilibria in large games. J Econ Theory 144:1300–1319
5. Fajardo S, Keisler HJ (2002) Model theory of stochastic processes. A. K. Peters Ltd, Massachusetts
6. Fremlin DH (2011) Measure theory: the irreducible minimum, vol 1. Torres Fremlin, Colchester
7. Fu HF (2008) Mixed-strategy equilibria and strong purification for games with private and public information. Econ Theor 37:521–432
8. Grant S, Meneghel I, Tourky R (2015) Savage games, Theoretical Economics, forthcoming
9. Greinecker M, Podczeck K (2015) Purification and roulette wheels. Econ Theor 58:255–272
10. Harsanyi JC (1983) Papers in game theory. D. Reidel Pub. Co., Dordrecht
11. Hoover D, Keisler HJ (1984) Adapted probability distributions. Trans Am Math Soc 286:159–201
12. He W, Sun YN (2013) The necessity of nowhere equivalence, working paper, National University of Singapore
13. He W, Sun X (2014) On the diffuseness of incomplete information game. J Math Econ 54:131–137
14. He W, Sun X, Sun YN (2013) Modeling infinitely many agents, working paper, National University of Singapore. (Revised version, June 28, 2015. Forthcoming Theoretical Economics)
15. Kakutani S (1944) Construction of a non-separable extension of the Lebesque measure space. Proc Imp Acad 20:115–119
16. Keisler HJ, Sun YN (2009) Why saturated probability spaces are necessary. Adv Math 221:1584–1607
17. Khan M. Ali, Rath KP, Sun YN (1999) On a private information game without pure strategy equilibria. J Math Econ 31:341–359
18. Khan M. Ali, Rath KP, Sun YN (2006) The Dvoretzky-Wald-Wolfowitz theorem and purification in atomless finite-action games. Int J Game Theory 34:91–104
19. Khan M. Ali, Rath KP, Sun YN, Yu H (2013) Large games with a bio-social typology. J Econ Theory 148:1122–1149
20. Khan M. Ali, Rath KP, Sun YN, Yu H (2015) Strategic uncertainty and the ex-post Nash property in large games. Theor Econ 10:103–129
21. Khan M. Ali, Rath KP, Yu H, Zhang Y (2013) Large distributional games with traits. Econ Lett 118:502–505
22. Khan M. Ali, Rath KP, Yu H, Zhang Y (2014) Strategic representation and realization of large distributional games, Johns Hopkins University, mimeo. An earlier version presented at the Midwest Economic Theory Group Meetings held in Lawrence, Kansas, 14–16 October 2005
23. Khan M. Ali, Sagara N (2013) Maharam-types and Lyapunov's theorem for vector measures on Banach spaces. Ill J Math 57:145–169

24. Khan M. Ali, Sun YN (1996) Nonatomic games on Loeb spaces. Proc Nat Acad Sci USA 93, 15518–15521
25. Khan M. Ali, Sun YN (2002) Non-cooperative games with many players. In: Aumann RJ, Hart S (eds) Handbook of game theory, vol 3. Elsevier Science, Amsterdam, pp 1761–1808 Chapter 46
26. Khan M. Ali, Zhang YC (2012) Set-valued functions, Lebesgue extensions and saturated probability spaces. Adv Math 229:1080–1103
27. Khan M. Ali, Zhang YC (2014) On the existence of pure-strategy equilibria in games with private information: a complete characterization. J Math Econ 50:197–202
28. Khan M. Ali, Zhang YC (2017) Existence of pure-strategy equilibria in Bayesian games: A sharpened necessity result. Int J Game Theory 46:167–183
29. Khan M. Ali, Zhang YC (2014) On pure-strategy equilibria in games with correlated information. Johns Hopkins University, mimeo
30. Loeb PA (1975) Conversion from nonstandard to standard measure spaces and applications in probability theory. Trans Amer Math Soc 211:113–122
31. Loeb PA, Sun YN (2009) Purification and saturation. Proc Am Math Soc 137:2719–2724
32. Mordukhovich BS, Sagara N, Subdifferentials of nonconvex integral functionals in Banach spaces with applications to stochastic dynamic programming. J Convex Anal, forthcoming. http://arxiv.org/abs/1508.02239
33. Maharam D (1942) On homogeneous measure algebras. Proc Nat Acad Sci USA 28:108–111
34. Milgrom PR, Weber RJ (1985) Distributional strategies for games with incomplete information. Math Oper Res 10:619–632
35. Myerson R (2004) Harsanyi's games with incomplete information. Manag Sci 50:1818–1824
36. Nash J (2007) The essential John Nash. In: Kuhn HW, Nassar S (eds). Princeton University Press, Princeton
37. Nillsen R (2010) Randomness and recurrence in dynamical systems, Carus Mathematical Monographs no 31. MAA Service Center, Washington
38. Podczeck K (2009) On purification of measure-valued maps. Econ Theor 38:399–418
39. Qiao L, Yu H (2014) On large strategic games with traits. J Econ Theory 153:177–190
40. Radner R, Rosenthal RW (1982) Private information and pure strategy equilibria. Math Oper Res 7:401–409
41. Radner R, Ray D (2003) Robert W. Rosenthal. J Econ Theory 112:365–368
42. Rashid S (1985) The approximate purification of mixed strategies with finite observation sets. Econ Lett 19:133–135
43. Reny P (2011) On the existence of monotone pure strategy equilibria in Bayesian games. Econometrica 79:499–553
44. Schmeidler D (1973) Equilibrium points of non-atomic games. J Stat Phys 7:295–300
45. Sun YN, Zhang YC (2009) Individual risk and Lebesgue extension without aggregate uncertainty. J Econ Theory 144:432–443
46. Wang J, Zhang YC (2012) Purification, saturation and the exact law of large numbers. Econ Theory 50:527–545

On Supermartingale Problems

Shigeo Kusuoka

Abstract In the present paper the author introduces a new notion, supermartingale problems, to describe a family of probability measures on path space, and shows some results on existence and stability.

Keywords Martingale problem · Supermartingale · Stochastic differential equation

Article type: Research Article
Received: January 17, 2017
Revised: January 27, 2017

1 Introduction

In mathematical finance, we usually think of a family of probability measures on path space (e.g. Equivalent Martingale Measures). In market models with no friction, these measures are equivalent and we can describe them by using a reference measure. However, in market models with friction, these measures are sometimes mutually singular (e.g. Boyle–Vorst [1], Kusuoka [4]). In the present paper we introduce a new notion, supermartingale problems, to describe such a family of probability measures and show some results on existence and stability.

Let $d \geqq 1$. We denote by W^d the space of continuous functions from $[0, \infty)$ to $\mathbf{R}^d$. Then W^d is a Polish space with a usual metric. We think of a filtration $\{\mathscr{F}_t^W\}_{t\in[0,\infty)}$ on W^d given by $\mathscr{F}_t^W = \sigma\{w(s); s \leqq t\}$, $t \geqq 0$. Also, we denote the space of probability measures on W^d by $\mathscr{P}(W^d)$. Then $\mathscr{P}(W^d)$ is a Polish space with the Prohorov metric. Also, we denote by S^d the set of $d \times d$ symmetric real matrices, and we

JEL Classification: C65, G00.
Mathematics Subject Classification (2010): 60G99, 60H99.

S. Kusuoka (✉)
Graduate School of Mathematical Sciences, The University of Tokyo,
Komaba 3-8-1, Meguro-ku, Tokyo 153-8914, Japan
e-mail: spfu5sh9@car.ocn.ne.jp

S. Kusuoka and T. Maruyama (eds.), *Advances in Mathematical Economics*, Advances in Mathematical Economics 21,
DOI 10.1007/978-981-10-4145-7_3

denote by S^d_+ the set of $d \times d$ non-negative definite symmetric real matrices. We regard S^d as a vector space with an inner product defined by $(A, B)_{S^d} = trace(AB)$, $A, B \in S^d$.

Definition 1 We say that $G : [0, \infty) \times W^d \times S^d \times \mathbf{R}^d \to \mathbf{R}$ is an HJ function, if the following are satisfied.

(1) $G(\cdot, \cdot, A, b) : [0, \infty) \times W^d \to \mathbf{R}$ is $\{\mathscr{F}_t^W\}$ progressively measurable for each $A \in S^d$ and $b \in \mathbf{R}^d$.
(2) $G(t, w, \cdot, \cdot) : S^d \times \mathbf{R}^d \to \mathbf{R}$ is convex and positive homogeneous for any $t \geqq 0$ and $w \in W^d$.
(3) $G(t, w, -A, 0) \leqq 0$ for any $A \in S^d_+$.

We denote by $\mathscr{H}\mathscr{J}^d$ the set of HJ functions $G : [0, \infty) \times W^d \times S^d \times \mathbf{R}^d \to \mathbf{R}$. For any $G \in \mathscr{H}\mathscr{J}^d$ and $f \in C^2(\mathbf{R}^d)$, we define $L_G f : [0, \infty) \times W^d \to \mathbf{R}$ by

$$L_G f(t, w) = G(t, w, \frac{1}{2}(\nabla^2 f)(w(t)), (\nabla f)(w(t)))$$

where

$$(\nabla^2 f)(x) = \left\{ \frac{\partial^2 f}{\partial x^i \partial x^j}(x) \right\}_{i,j=1,\ldots,d} \in \mathscr{S}^d$$

and

$$(\nabla f)(x) = \left(\frac{\partial f}{\partial x^i}(x) \right)_{i=1,\ldots,d} \in \mathbf{R}^d, \qquad x \in \mathbf{R}^d.$$

Definition 2 We say that μ is a solution to the supermartingale problem associated with the HJ function G, if μ is a probability measure on W^d satisfying the following.

(1) $E^\mu \left[\int_0^t |L_G f(s, w)| ds \right] < \infty$ for any $f \in C_0^\infty(\mathbf{R}^d)$ and $t > 0$.
(2) $\left\{ f(w(t)) - \int_0^t L_G f(s, w) ds;\ t \geqq 0 \right\}$ is a supermartingale under $\mu(dw)$ for any $f \in C_0^\infty(\mathbf{R}^d)$.

For any HJ function G and $x \in \mathbf{R}^N$, we denote by $\mathscr{R}(G, x)$ the set of solutions μ to the supermartingale problem associated with the HJ function G satisfying $\mu(w(0) = x) = 1$.

2 Preliminary Facts

Let E be a finite dimensional vector space with an inner product $(\cdot, \cdot)_E$. Also, let $|\cdot|_E : E \to [0, \infty)$ be a norm on E given by $|x|_E = (x, x)_E^{1/2}$, $x \in E$. Let $\mathscr{C}(E)$ denote the set of convex and positive homogeneous functions defined in E, and

$\mathcal{K}(E)$ denote the set of non-void compact convex sets in E. We define a map $\hat{K} : \mathcal{C}(E) \to \mathcal{K}(E)$ and $\Phi : \mathcal{K}(E) \to \mathcal{C}(E)$ by

$$\hat{K}(f) = \{\xi \in E;\ f(x) \geqq (x, \xi)_E \text{ for all } x \in E\}, \quad f \in \mathcal{C}(E),$$

and

$$\Phi(K)(x) = \sup\{(x, \xi)_E;\ \xi \in K\}, \quad x \in E,\ K \in \mathcal{K}(E).$$

Then we have the following.

Proposition 1 (1) *$\Phi \circ \hat{K}$ is the identity map in $\mathcal{C}(E)$.*
(2) *$\hat{K} \circ \Phi$ is the identity map in $\mathcal{K}(E)$.*

Proof (1) Let $f \in \mathcal{C}(E)$. It is obvious that $(\Phi \circ \hat{K})(f)(x) \leqq f(x)$, $x \in E$.

Let us take an arbitrary point x_0 in E. Since f is a convex function, there is a $\xi_0 \in E$ such that

$$f(x) \geqq f(x_0) + (x - x_0, \xi_0)_E, \qquad x \in E.$$

Since f is positive homogeneous, for any $a > 0$

$$af(x) = f(ax) \geqq f(x_0) + (ax - x_0, \xi_0)_E \geqq a(x, \xi_0)_E + f(x_0) - (x_0, \xi_0)_E, \qquad x \in E.$$

Letting $a \downarrow 0$ we have $0 \geqq f(x_0) - (x_0, \xi_0)_E$. Also, we see that

$$f(x) \geqq (x, \xi_0)_E + \frac{1}{a}(f(x_0) - (x_0, \xi_0)_E), \qquad x \in E.$$

Letting $a \to \infty$, we have

$$f(x) \geqq (x, \xi_0)_E \qquad x \in E.$$

This implies $\xi_0 \in \hat{K}(f)$. So we see that

$$f(x_0) \leqq (x_0, \xi_0)_E \leqq (\Phi \circ \hat{K})(f)(x_0).$$

Since x_0 is arbitrary, we have Assertion (1).

(2) Let $K \in \mathcal{K}(E)$. It is obvious that if $\eta \in K$, then $(x, \eta)_E \leqq \Phi(K)(x)$, $x \in E$. On the other hand, since K is a nonvoid compact convex set, if $\eta \notin K$, there is a $x_0 \in E$ such that

$$(x_0, \eta)_E > \sup\{(x_0, \xi)_E;\ \xi \in K\} = \Phi(K)(x_0)$$

These implies $(\hat{K} \circ \Phi)(K) = K$. ∎

The following is an easy consequence of the previous proposition.

Proposition 2 *For any $f_1, f_2 \in \mathscr{C}(E)$, $\hat{K}(f_1) \subset \hat{K}(f_2)$, if and only if $f_1(x) \leqq f_2(x)$ for all $x \in E$. In particular, for any $f \in \mathscr{C}(E)$ and $r > 0$, $\hat{K}(f) \subset \{\xi \in E;\ |\xi|_E \leqq r\}$, if and only if $f(x) \leqq r|x|_E$ for all $x \in E$.*

It is easy to see that for any $K \in \mathscr{K}(E)$ and $x \in E$, there is a unique $y \in K$ such that $|x - y|_E = d_E(x, K)$. We denote by $P(x, K)$ this point y.

Proposition 3 *Let $K \in \mathscr{K}(E)$ and $x \in E$. Then we see that*

$$(x - P(x, K), z - P(x, K))_E \leqq 0 \text{ for any } z \in K.$$

Proof If $x \in K$, the assertion is obvious. So we assume that $x \notin K$.

For any $z \in K$, let $z(t) = tz + (1 - t)P(x, K)$, $t \in (0, 1)$. Then $z(t) \in K$, $t \in (0, 1)$, and

$$|x - P(x, K)|_E^2 \leqq |x - z(t)|_E^2 = |x - P(x, K) + t(P(x, K) - z)|_E^2$$

$$= |x - P(x, K)|_E^2 + t^2|P(x, K) - z)|_E^2 - 2t(x - P(x; K), z - P(x; K))_E$$

for $\in (0, 1)$. So we have

$$t|P(x, K) - z)|_E^2 - 2(x - P(x; K), z - P(x; K))_E \geqq 0, \qquad t \in (0, 1).$$

This implies our assertion. ■

Now let us think of the Hausdorff metric d_H^E on $\mathscr{K}(E)$, i.e.,

$$d_H^E(K, K') = \max\{\sup\{d_E(\xi, K); \xi \in K'\}, \sup\{d_E(\xi, K'); \xi \in K\}\}$$

for $K, K' \in \mathscr{K}(E)$. Then we have the following.

Proposition 4 *For any $K, K' \in \mathscr{K}(E)$,*

$$d_H^E(K, K') = \sup\{|\Phi(K)(x) - \Phi(K')(x)|;\ x \in E \text{ with } |x|_E \leqq 1\}.$$

Proof Let $r = d_H(K, K')$ and $r' = \sup\{|\Phi(K)(x) - \Phi(K')(x)|; x \in E$ with $|x|_E \leqq 1\}$. It is sufficient to think of the case that $r > 0$.

First, we show that $r \leqq r'$. We may assume that $\sup\{d_E(\xi, K); \xi \in K'\} = r$ without loss of generality. Since K' is compact, there is a $\xi_0 \in K'$ such that $d_E(\xi_0, K) = r$. So letting $x_0 = r^{-1}(\xi_0 - P(\xi_0, K)))$, we see by Proposition 3 that $|x_0|_E = 1$ and $(x_0, \eta - P(\xi_0, K))_E \leqq 0$, $\eta \in K$. So we have

$$\Phi(K)(x_0) \leqq (x_0, P(\xi_0, K))_E = (x_0, \xi_0)_E - r \leqq \Phi(K')(x_0) - r$$

So we see that $r \leqq r'$.

Next we show that $r' \leqq r$. Let $x \in E$ with $|x|_E = 1$. Then there is a $\xi \in K$ such that $\Phi(K)(x) = (x, \xi)_E$. Since there is an $\eta \in K'$ such that $|\xi - \eta|_E \leqq r$, we see that $\Phi(K)(x) \leqq (x, \eta)_E + r \leqq \Phi(K')(x) + r$. So we have $\sup\{\Phi(K)(x) - \Phi(K')(x);\ x \in E, |x|_E = 1\} \leqq r$. This observation shows that $r' \leqq r$.

This completes the proof. ■

Proposition 5 *For any $K \in \mathscr{K}(E)$ and $x_0, x_1 \in E$,*

$$|P(x_1, K) - P(x_0, K)|_E \leqq 2|x_1 - x_0|_E.$$

Proof Let $y_0 = P(x_0, K)$ and $y_1 = P(x_1, K)$. Then by Proposition 3 we see that $(x_0 - y_0, y_1 - y_0)_E \leqq 0$. Since $(x_1 - y_1, x_1 - y_1)_E \leqq (x_1 - y_0, x_1 - y_0)_E$, we see that

$$(y_0 - y_1, y_0 - y_1)_E = |x_1 - y_1|_E^2 - |x_1 - y_0|_E^2 - 2(y_0 - y_1, x_1 - y_0)_E$$

$$\leqq 2(y_1 - y_0, x_0 - y_0)_E + 2(y_1 - y_0, x_1 - x_0)_E \leqq 2(y_1 - y_0, x_1 - x_0)_E.$$

Then we have $|y_1 - y_0|_E^2 \leqq 2|y_1 - y_0|_E|x_1 - x_0|_E$. This implies our assertion. ■

Proposition 6 *For any $K, K' \in \mathscr{K}(E)$, and $x \in E$,*

$$|P(x, K) - P(x, K')|_E$$

$$\leqq 2d_H(K, K')^{1/2}(d_H(K, K') + 2(d_E(x, K) \vee d_E(x, K')))$$

Proof We may assume that $d(x, K') \leqq d(x, K)$ without loss of generality. Let $r = d_H(K, K')$, $y = P(x, K)$, and $y' = P(x, K')$. Then we see that $|y' - x|_E \leqq |y - x|_E$. Note that there is a $z \in K$ such that $|y' - z|_E \leqq r$. Then by Proposition 3 we see that that $(z - y, x - y)_E \leqq 0$. Therefore we see that

$$|z - x|_E^2 = |y - x|_E^2 + |z - y|_E^2 - 2(z - y, x - y)_E \geqq |y - x|_E^2 + |z - y|_E^2.$$

So we have

$$|z - y|_E^2 \leqq (|y' - x|_E + |z - y'|_E)^2 - |y - x|_E^2$$

$$\leqq (|y' - x|_E + r)^2 - |y - x|_E^2 \leqq r(r + 2|y' - x|_E).$$

So we have

$$|y - y'|_E \leqq |y' - z|_E + |z - y|_E \leqq r + r^{1/2}(r + 2d_E(x, K'))^{1/2}$$

$$\leqq 2r^{1/2}(r + 2d_E(x, K))^{1/2}$$

This implies our assertion. ■

Now let us assume that dim $E = N$. Let $\mathscr{K}_+(E) = \{K \in \mathscr{K}(E);\ \nu_E(K) > 0\}$. Here ν_E is the volume measure in E.

Let $\hat{c} : \mathscr{K}_+(E) \to E$ be given by

$$\hat{c}(K) = \frac{1}{\nu_E(K)} \int_K x \nu_E(dx) \qquad K \in \mathscr{K}_+(E).$$

It is easy to see that $\hat{c}(K) \in K$ for any $K \in \mathscr{K}_+(E)$. Also, it is obvious that $\hat{c}(K - \hat{c}(K)) = 0$.

Proposition 7 *Let $K \in \mathscr{K}_+(E)$ and assume that $\hat{c}(K) = 0$. Then $\Phi(K)(x) > 0$ for any $x \in E \setminus \{0\}$.*

Proof Let $x \in E \setminus \{0\}$. Since $0 \in K$, it is obvious that $\Phi(K)(x) \geqq 0$. Suppose that $\Phi(K)(x) = 0$. Then we see that $(x, z)_E \leqq 0$ for any $z \in K$. However, $\nu_E(\{z \in E;\ (x, z)_E = 0\}) = 0$. So we see that $(x, z)_E < 0\ \nu_E - a.e. z \in K$. This implies that

$$0 = \int_K (x, z)_E \nu_E(dz) < 0.$$

This is a contradiction. So we see that $\Phi(K)(x) > 0$. ■

For any $R > 0$, let $B_E(x, R) = \{y \in E;\ |y - x|_E \leqq R\}$, $x \in E$.

Proposition 8 *Let $K \in \mathscr{K}_+(E)$ with $\hat{c}(K) = 0$. Let*

$$r_0(K) = \min\{\Phi(K)(x)\ ; x \in E,\ |x|_E = 1\}$$

and

$$r_1(K) = \max\{\Phi(K)(x)\ ; x \in E,\ |x|_E = 1\}$$

Then $r_0(K) > 0$ and $B_E(0, r_0(K)) \subset K \subset B_E(0, r_1(K))$.

In particular, $v_N r_0(K)^N \leqq \nu_E(K) \leqq v_N r_1(K)^N$, where $v_N = \nu_E(B(0, 1))$ and depends only on $N = \dim E$.

Proof We see that $r_0(E) > 0$ by Proposition 7. Note that

$$B_E(0, R) = \{\xi \in E;\ R|x|_E \geqq (x, \xi)_E \text{ for all } x \in E\}.$$

So we see that

$$B_E(0, r_0(K)) \subset \hat{K}(\Phi(K)) \subset B_E(0, r_1(K)).$$

Therefore we have our assertion from Proposition 1. ■

Proposition 9 *Let $R_1 > R_0 > 0$, and $K \in \mathscr{K}(E)$ and assume that $B_E(0, R_0) \subset K \subset B_E(0, R_1)$. Then*

$$B_E(\hat{c}(K), \left(\frac{R_0}{R_1}\right)^N R_0) \subset K.$$

Proof For any $x \in E$, we have

$$\begin{aligned}
(x, \hat{c}(K))_E &= \frac{1}{\nu_E(K)} \int_{K \setminus B_E(0, R_0)} (x, \xi)_E \nu_E(d\xi) \\
&\leqq \frac{\nu_E(K \setminus B_E(0, R_0))}{\nu_E(K)} \Phi(K)(x) \leqq \left(1 - \frac{\nu_E(B_E(0, R_0))}{\nu_E(B_E(0, R_1))}\right) \Phi(K)(x) \\
&= \left(1 - \left(\frac{R_0}{R_1}\right)^N\right) \Phi(K)(x).
\end{aligned}$$

Since $B_E(0, R_0) \subset K$, we see that $\Phi(K)(x) \geqq R_0 |x|_E$. So if $\eta \in B_E(\hat{c}(K), \left(\frac{R_0}{R_1}\right)^N R_0)$, then for any $x \in E$,

$$\begin{aligned}
(x, \eta)_E &= (x, \eta - \hat{c}(K))_E + (x, \hat{c}(K))_E \\
&\leqq \left(\frac{R_0}{R_1}\right)^N R_0 |x|_E + \left(1 - \left(\frac{R_0}{R_1}\right)^N\right) \Phi(K)(x) \leqq \Phi(K)(x).
\end{aligned}$$

So by Proposition 1, we have $\eta \in \hat{K}(\Phi(K)) = K$. So we have our assertion. ■

Now for any $R_1 > R_0 > 0$, let

$$\begin{aligned}
&\mathscr{K}_{R_0, R_1}(E) \\
&= \{K \in \mathscr{K}(E);\ \text{there is an } x \in E \text{ such that } B_E(x, R_0) \subset K \subset B_E(x, R_1)\}.
\end{aligned}$$

Then by Proposition 9, we have the following.

Proposition 10 *Let $R_1 > R_0 > 0$. Then for any $K \in \mathscr{K}_{R_0, R_1}(E)$,*

$$B_E(\hat{c}(K), \left(\frac{R_0}{R_1}\right)^N R_0) \subset K.$$

Proposition 11 *Let $R_1 > R_0 > 0$. Then for any $K_0, K_1 \in \mathscr{K}_{R_0, R_1}(E)$,*

$$|\hat{c}(K_1) - \hat{c}(K_0)|_E \leqq 2^{N+4} \left(\frac{R_1}{R_0}\right)^{N+1} d_H^E(K_1, K_0).$$

Proof Let $r = d_H(K_1, K_0)$. There is an $x_0 \in E$ such that $B_E(x_0, R_0) \subset K_0 \subset B_E(x_0, R_1)$.

If $x \in K_0 \cup K_1$, there is a $y \in K_0$ such that $|x - y|_E \leqq r$. Then we see that

$$\frac{R_0}{R_0 + r}x + \frac{r}{R_0 + r}x_0 = \frac{R_0}{R_0 + r}y + \frac{r}{R_0 + r}\left(\frac{R_0}{r}(x - y) + x_0\right) \in K_0.$$

So we see that

$$K_0 \cup K_1 \subset \frac{R_0 + r}{R_0}\left(K_0 - \frac{r}{R_0 + r}x_0\right).$$

So we have

$$\nu_E(K_1 \setminus K_0) \leqq \nu_E\left(\frac{R_0 + r}{R_0}\left(K_0 - \frac{r}{R_0 + r}x_0\right)\right) - \nu_E(K_0)$$
$$= \left(\left(\frac{R_0 + r}{R_0}\right)^N - 1\right)\nu_E(K_0) \leqq \left(\left(1 + \frac{r}{R_0}\right)^N - 1\right)v_N R_1^N.$$

Similarly we have

$$\nu_E(K_0 \setminus K_1) \leqq \left(\left(\frac{R_0 + r}{R_0}\right)^N - 1\right)v_N R_1^N$$

and so we have

$$\nu_E((K_0 \cup K_1) \setminus (K_0 \cap K_1)) \leqq 2\left(\left(1 + \frac{r}{R_0}\right)^N - 1\right)v_N R_1^N$$

Also, we see that

$$\nu_E((K_0 \cup K_1) \setminus (K_0 \cap K_1)) \leqq \nu_E(K_0) + \nu_E(K_1) \leqq 2v_N R_1^N.$$

So we have

$$|\nu_E(K_0) - \nu_E(K_1)| \leqq \nu_E((K_0 \cup K_1) \setminus (K_0 \cap K_1))$$
$$\leqq 2\left(\left(\left(1 + \frac{r}{R_0}\right) \wedge 2\right)^N - 1\right)v_N R_1^N \leqq 2^{N+1}\frac{r}{R_0}v_N R_1^N.$$

Note that $K_1 \subset B_E(x_0, R_1 + r)$. So we have

$$|\hat{c}(K_1) - \hat{c}(K_0)|_E$$

$$
\begin{aligned}
&= |\frac{1}{\nu_E(K_1)}\int_{K_1}(x-x_0)\nu_E(dx) - \frac{1}{\nu_E(K_0)}\int_{K_0}(x-x_0)\nu_E(dx)|_E \\
&\leqq |\frac{1}{\nu_E(K_1)}\left(\int_{K_1}(x-x_0)\nu_E(dx) - \int_{K_0}(x-x_0)\nu_E(dx)\right)|_E \\
&\qquad + |\frac{1}{\nu_E(K_1)} - \frac{1}{\nu_E(K_0)}||\int_{K_0}(x-x_0)\nu_E(dx)|_E \\
&\leqq \frac{\nu_E((K_0\cup K_1)\setminus(K_0\cap K_1))}{\nu_E(K_1)}(R_1+r) + \frac{|\nu_E(K_1)-\nu_E(K_0)|}{\nu_E(K_1)}R_1 \\
&\leqq 2^{N+2}\frac{r}{R_0}\left(\frac{R_1}{R_0}\right)^N (R_1+r).
\end{aligned}
$$

Since there is a $z_0 \in K_0$, such that $|\hat{c}(K_1) - z_0|_E \leqq r$, we see that

$$|\hat{c}(K_1) - \hat{c}(K_0)|_E \leqq |\hat{c}(K_1) - z_0|_E + |z_0 - x_0|_E + |\hat{c}(K_0) - x_0|_E \leqq r + 2R_1.$$

Combining them, we have our assertion. ■

Let $\Psi : E \times \mathscr{K}_+(E) \to E$ be given by

$$\Psi(x, K) = \hat{c}(K) + \max\{t \in [0,1];\ t(x - \hat{c}(K)) + \hat{c}(K) \in K\}(x - \hat{c}(K)) \quad (1)$$

for $x \in E$ and $K \in \mathscr{K}_+(E)$.

Then we have the following.

Proposition 12 *Let $R_1 > R_0 > 0$. Then for any $K \in \mathscr{K}_{R_0,R_1}(E)$, and $x, y \in E$,*

$$|\Psi(x, K) - \Psi(y, K)|_E \leqq (1 + 2\left(\frac{R_1}{R_0}\right)^{N+1})|x - y|_E.$$

Proof Let $R_2 = (R_0/R_1)^N R_0$. Then by Proposition 10,

$$B_E(\hat{c}(K), R_2) \subset K \subset B_E(\hat{c}(K), 2R_1).$$

We may assume $\hat{c}(K) = 0$ without loss of generality.

Let $f : E \to [0, 1]$ be given by

$$f(x) = \max\{t \in [0,1];\ tx \in K\}.$$

Assume that $x, y \in E$, $x \neq y$, and $f(y) \leqq f(x)$. Then we see that $f(x)x \in K$, and that

$$f(x)R_2\frac{1}{|x-y|_E}(y-x)\in K.$$

So we have

$$f(x)\frac{R_2}{R_2+|x-y|_E}y$$
$$=f(x)\frac{R_2}{R_2+|x-y|_E}x+\frac{|x-y|_E}{R_2+|x-y|_E}f(x)R_2\frac{1}{|x-y|_E}(y-x)\in K.$$

So we have

$$f(y)\geqq f(x)\frac{R_2}{R_2+|x-y|_E}.$$

Therefore we have

$$f(x)-f(y)\leqq f(x)\frac{|x-y|_E}{R_2+|x-y|_E}.$$

Then

$$|\Psi(x,K)-\Psi(y,K)|_E=|f(x)x-f(y)y|_E\leqq(f(x)-f(y))|x|_E+f(y)|x-y|_E$$
$$\leqq\frac{|x-y|_E}{R_2+|x-y|_E}f(x)|x|_E+|x-y|_E\leqq\left(1+\frac{2R_1}{R_2}\right)|x-y|_E.$$

This implies our assertion. ■

Proposition 13 *Let* $R_1>R_0>0$. *Let* $K_0,K_1\in\mathscr{K}_{R_0,R_1}(E)$ *and assume that* $\hat{c}(K_1)=\hat{c}(K_0)=0$. *Then for any* $x\in E$,

$$|\Psi(x,K_1)-\Psi(x,K_0)|_E\leqq 2\left(\frac{R_1}{R_0}\right)^{N+1}d_H^E(K_1,K_0).$$

Proof Let $r=d_H^E(K_1,K_0)$, and

$$t_i=\max\{s\in[0,1];\ sx\in K_i\},\qquad i=0,1.$$

Since $t_0x\in K_0$, we see that there is a $y\in K_1$ such that $|t_0x-y|_E\leqq r$.

Let $R_2=(R_0/R_1)^NR_0$. Then by Proposition 10, we see that $B_E(0,R_2)\subset K_1\subset B_E(0,2R_1)$, and so $R_2r^{-1}(t_0x-y)\in K_1$. Therefore

$$t_0\frac{R_2}{R_2+r}x=\frac{R_2}{R_2+r}y+\frac{r}{R_2+r}\frac{R_2}{r}(t_0x-y)\in K_1.$$

This implies

$$t_1 \geqq t_0 \frac{R_2}{R_2 + r},$$

and so

$$t_0 - t_1 \leqq \frac{r}{R_2 + r} t_0 \leqq \frac{r}{R_2 + r} \left(1 \wedge \frac{2R_1}{|x|_E}\right).$$

Similarly we have

$$t_1 - t_0 \leqq \frac{r}{R_2 + r} \left(1 \wedge \frac{2R_1}{|x|_E}\right).$$

Therefore we have

$$|\Psi(x, K_1) - \Psi(x, K_0)|_E = |t_1 - t_0||x|_E \leqq \frac{2R_1}{R_2} r.$$

This implies our assertion. ■

Theorem 1 *Let $R_1 > R_0 > 0$. Then there is a $C_{R_0,R_1} < \infty$ such that*

$$|\Psi(x_1, K_1) - \Psi(x_0, K_0)|_E \leqq C_{R_0,R_1}(|x_1 - x_0|_E + d_H^E(K_1, K_0))$$

for any $K_0, K_1 \in \mathscr{K}_{R_0,R_1}(E)$, and $x_0, x_1 \in E$.

Proof Note that

$$\Psi(x_i, K_i) = \Psi(x_i - \hat{c}(K_i), K_i - \hat{c}(K_i)) + \hat{c}(K_i), \qquad i = 0, 1.$$

So by Propositions 12 and 13 we have

$$\begin{aligned}
&|\Psi(x_1, K_1) - \Psi(x_0, K_0)|_E \\
&\leqq |\Psi(x_1 - \hat{c}(K_1), K_1 - \hat{c}(K_1)) - \Psi(x_0 - \hat{c}(K_0), K_0 - \hat{c}(K_0))|_E + |\hat{c}(K_1)) - \hat{c}(K_0))|_E \\
&\leqq |\Psi(x_1 - \hat{c}(K_1), K_1 - \hat{c}(K_1)) - \Psi(x_0 - \hat{c}(K_0), K_1 - \hat{c}(K_1))|_E \\
&\quad + |\Psi(x_0 - \hat{c}(K_0), K_1 - \hat{c}(K_1)) - \Psi(x_0 - \hat{c}(K_0), K_0 - \hat{c}(K_0))|_E + |\hat{c}(K_1)) - \hat{c}(K_0))|_E \\
&\leqq \left(1 + 2\left(\frac{R_1}{R_0}\right)^{N+1}\right) |(x_1 - \hat{c}(K_1)) - (x_0 - \hat{c}(K_0))|_E \\
&\quad + 2\left(\frac{R_1}{R_0}\right)^{N+1} d_H^E(K_1 - \hat{c}(K_1), K_0 - \hat{c}(K_0)) + |\hat{c}(K_1) - \hat{c}(K_0))|_E.
\end{aligned}$$

Note that

$$\begin{aligned}
&d_E^H(K_1 - y_1, K_0 - y_0) \\
&= \max\{\sup\{d_E(z - y_1, K_0 - y_0);\ z \in K_1\}, \sup\{d_E(z - y_0, K_1 - y_1);\ z \in K_0\}\} \\
&\leqq \max\{\sup\{d_E(z, K_0) + |y_1 - y_0|_E;\ z \in K_1\}, \sup\{d_E(z, K_1) + |y_1 - y_0|_E;\ z \in K_0\}\} \\
&= d_H^E(K_1, K_0) + |y_1 - y_0|_E.
\end{aligned}$$

This observation shows that

$$d_H^E(K_1 - \hat{c}(K_1), K_0 - \hat{c}(K_0)) \leqq d_H^E(K_1, K_0) + |\hat{c}(K_1) - \hat{c}(K_0))|_E.$$

So by Proposition 11 we have our assertion. ■

Example Let $K_\theta \in \mathcal{K}(\mathbf{R}^2)$, $\theta \in [0, \pi/4]$, given by

$$K_\theta = \{(x, y) \in [0, 2] \times [0, 1];\ x \geqq \cos^2\theta,\ (\sin\theta)y \geqq \cos\theta(1 - x)\}, \qquad \theta \in [0, \frac{\pi}{4}].$$

Then it is easy to see that $P((0, 0), K_\theta) = (\cos^2\theta, \sin\theta\cos\theta)$, and $d_H(K_\theta, K_0) = 1 - \cos^2\theta$. So we see that

$$\lim_{\theta\downarrow 0} \frac{|P((0, 0), K_\theta) - P((0, 0), K_0)|^2}{d_H(K_\theta, K_0)} = 1$$

On the other hand, we have by Theorem 1 we see that

$$\overline{\lim_{\theta\downarrow 0}} \frac{|\Psi((0, 0), K_\theta) - \Psi((0, 0), K_0)|}{d_H(K_\theta, K_0)} < \infty.$$

3 Square Root

Proposition 14 *Let $A \in S_+^d$. Then we have*

$$||(A + sI_d)^{1/2} - A^{1/2}||_{S^d} \leqq d^{1/2}s^{1/2}, \qquad s \geqq 0.$$

Here I_d denotes the $d \times d$ identity matrix.

Proof Let λ_k, $k = 1, \ldots, d$ be the eigen values of A. Then we have

$$||(A + sI_d)^{1/2} - A^{1/2}||_{S^d}^2 = \sum_{k=1}^d ((\lambda_k + s)^{1/2} - \lambda_k^{1/2})^2 = \sum_{k=1}^d \left(\int_{\lambda_k}^{\lambda_k+s} \frac{1}{2}x^{-1/2}dx\right)^2$$

$$\leqq \sum_{k=1}^{d} \left(\int_0^s \frac{1}{2} x^{-1/2} dx \right)^2 = ds.$$

This implies our assertion. ∎

Proposition 15 (1) *Let* $A, B \in S_+^d$, *and assume that the minimum eigenvalue* $\lambda_{min}(A)$ *of* A *is positive. Then if* $||B - A||_{S^d} \leqq \lambda_{min}(A)/2$, *we have*

$$||B^{1/2} - A^{1/2}||_{S^d} \leqq \lambda_{min}(A)^{-1/2} ||B - A||_{S^d}.$$

(2) *For any* $A, B \in S_+^d$,

$$||B^{1/2} - A^{1/2}||_{S^d} \leqq 4d^{1/2} ||B - A||_{S^d}^{1/2}.$$

(3) *Let* $A, B \in S_+^d$, *and assume that the minimum eigenvalue* $\lambda_{min}(A)$ *of* A *is positive. Then we have*

$$||B^{1/2} - A^{1/2}||_{S^d} \leqq 8d^{1/2} \lambda_{min}(A)^{-1/2} ||B - A||_{S^d}.$$

Proof First we prove Assertion (1). Let $c = \lambda_{min}(A)^{1/2}$, and let $F : S^d \to S^d$ be a linear operator given by

$$F(D) = A^{1/2} D + D A^{1/2}, \qquad D \in S^d.$$

If A is diagonal and is given by $A = (\lambda_i \delta_{ij})_{i,j=1,\ldots,d}$, then $F(D) = ((\lambda_i^{1/2} + \lambda_j^{1/2}) d_{ij})_{i,j=1,\ldots,d}$ for $D = (d_{ij})_{i,j=1,\ldots,d} \in S^d$. This observation shows that F is bijective and

$$||F^{-1}(D)||_{S^d} \leqq (2c)^{-1} ||D||_{S^d} \qquad D \in S^d.$$

Let $B \in S^d$ satisfy $||B - A||_{S^d} \leqq c^2/2$, and let us define $D_n \in S^d$, $n = 0, 1, \ldots,$ inductively by $D_0 = 0$ and

$$D_n = F^{-1}(B - A - D_{n-1}^2), \qquad n = 1, 2, \ldots$$

Then we see that

$$||D_n||_{S^d} \leqq (2c)^{-1} (||B - A||_{S^d} + ||D_{n-1}||_{S^d}^2) \leqq c/4 + (2c)^{-1} ||D_{n-1}||_{S^d}^2, \quad n = 1, 2, \ldots$$

In particular, we see that $||D_n|| \leqq c/2$, $n \geqq 0$, inductively. Then we see that

$$||D_{n+1} - D_n||_{S^d} \leqq (2c)^{-1} ||D_n^2 - D_{n-1}^2||_{S^d}$$

$$= (2c)^{-1} ||(D_n - D_{n-1}) D_n + D_{n-1} (D_n - D_{n-1})||_{S^d}$$

$$\leqq (2c)^{-1}(||D_n||_{S^d} + ||D_{n-1}||_{S^d})||D_n - D_{n-1}||_{S^d} \leqq \frac{1}{2}||D_n - D_{n-1}||_{S^d}, \qquad n = 1, 2, \ldots$$

So we see that

$$||D_n - D_{n-1}||_{S^d} \leqq (\frac{1}{2})^{n-1}(2c)^{-1}||B - A||_{S^d}$$

Therefore we see that there is a $D_\infty \in S^d$ such that $D_n \to D_\infty, n \to \infty$, and $||D_\infty||_{S^d} \leqq c^{-1}||B - A||_{S^d}$. Since $F(D_\infty) = (B - A) - D_\infty^2$, we see that $(D_\infty + A^{1/2})^2 = B$. Noting that $||D_\infty||_{S^d} \leqq c/2$, we see that $D_\infty + A^{1/2} \in S_+^d$. This implies that $D_\infty = B^{1/2} - A^{1/2}$. So we have Assertion (1).

Now let us prove Assertion (2). Let $r = ||B - A||_{S^d}$. Note that $r \leqq (\lambda_{min}(A) + 2r)/2$. So we have by (1)

$$||(B + 2rI_d)^{1/2} - (A + 2rI_d)^{1/2}||_{S^d} \leqq (\lambda_{min}(A) + 2r)^{-1/2}r \leqq r^{1/2}$$

Then by Proposition 14 we have

$$\begin{aligned}
&||B^{1/2} - A^{1/2}||_{S^d} \\
&\leqq ||(B + 2rI_d)^{1/2} - (A + 2rI_d)^{1/2}||_{S^d} + ||(B + 2rI_d)^{1/2} - B^{1/2}||_{S^d} \\
&\qquad\qquad + ||(A + 2rI_d)^{1/2} - A^{1/2}||_{S^d} \\
&\leqq 2d^{1/2}(2r)^{1/2} + r^{1/2} \leqq 4d^{1/2}r^{1/2}
\end{aligned}$$

This implies Assertion (2).

Assertion (3) is an easy consequence of Assertions (1) and (2). ■

4 Existence Theorem

For any probability measure μ on W^d, let $\{\mathscr{F}_t^\mu\}_{t\in[0,\infty)}$ be a filtration over W^d given by

$$\mathscr{F}_t^\mu = \bigcap_{s>t} \sigma\{\mathscr{F}_s^W \cup \mathscr{N}^\mu\}, \qquad t \in [0, \infty),$$

where

$$\mathscr{N}^\mu = \{B \subset W^d;\ \text{there exists an } A \in \mathscr{B}(W^d) \text{ such that } B \subset A \text{ and } \mu(A) = 0\}.$$

Let E^d denote $S^d \times \mathbf{R}^d$. Then E^d has an inner product $(\cdot, \cdot)_{E^d}$ given by

$$((A, a), (B, b))_{E^d} = \text{trace}(AB) + (a, b)_{\mathbf{R}^d}, \qquad A, B \in S^d,\ a, b \in \mathbf{R}^d.$$

For $G \in \mathscr{H}\mathscr{J}^d$, let $K_G : [0, \infty) \times W^d \to \mathscr{K}(E^d)$ be a progressively measurable map given by $K_G(t, w) = \hat{K}(G(t, w, \cdot, \cdot))$, $(t, w) \in [0, \infty) \times W^d$.

Then we have the following.

Proposition 16 *Let* $G \in \mathscr{H}\mathscr{J}^d$. *Then* $K_G(t, w) \subset (S^d_+ \times \mathbf{R}^d)$ *for any* $(t, w) \in [0, \infty) \times W^d$.

Proof Suppose that $(A, b) \in E^d \setminus S^d_+ \times \mathbf{R}^d$. Then there is a $B \in S^d_+$ such that $(A, B)_{S^d} < 0$. So we have

$$0 < ((A, b), (-B, 0))_{E^d} \leqq G(t, w, -B, 0).$$

This contradicts to the definition of HJ functions (Definition 1(3)). So we have our assertion. ■

Also, we have the following.

Proposition 17 *Let* $G \in \mathscr{H}\mathscr{J}^d$ *and* $x_0 \in \mathbf{R}^d$, *and assume that there is a constant* $C > 0$ *such that* $|G(t, w, z)| \leqq C|z|_{E^d}$ *for any* $(t, w, z) \in [0, \infty) \times W^d \times E^d$. *Then we have the following.*

(1) *Suppose that* $\mu \in \mathscr{R}(G, x_0)$. *Then* $\{w^i(t), t \geqq 0\}$, $i = 1, \ldots, d$, *under* μ *are* $\mathscr{F}^\mu_t$-*semi-martingales. Moreover, there are* $\{\mathscr{F}^\mu_t\}_{t \geqq 0}$-*progressively measurable processs* $a^{ij} : [0, \infty) \times W^d \to \mathbf{R}$, $i, j = 1, \ldots, d$, *and* $b^i : [0, \infty) \times W^d \to \mathbf{R}$, $i = 1, \ldots, d$, *such that* $w^i(t) - \int_0^t b^i(s)ds$, $t \geqq 0$, $i = 1, \ldots, d$, *are* $\mathscr{F}^\mu_t$-*martingales,*

$$\langle w^i, w^j \rangle_t = \int_0^t a^{ij}(s)ds, \; i, j = 1, \ldots, d, \qquad t \geqq 0, \; \mu - a.s.w, \tag{2}$$

and

$$(\{a^{ij}(t)\}_{i,j=1,\ldots,d}, (b^i(t))_{i=1,\ldots,d}) \in K_G(t, w) \qquad t \geqq 0, \; \mu - a.s.w. \tag{3}$$

(2) *Let* μ *be a probability measure on* W^d. *Suppose that* $\mu(w(0) = x_0) = 1$ *and that* $\{w^i(t), t \geqq 0\}$, $i = 1, \ldots, d$, *under* μ *are* $\mathscr{F}^\mu_t$-*semi-martingales. Suppose moreover that there are* $\{\mathscr{F}^\mu_t\}_{t \geqq 0}$-*progressively measurable processs* $a^{ij} : [0, \infty) \times W^d \to \mathbf{R}$, $i, j = 1, \ldots, d$, *and* $b^i : [0, \infty) \times W^d \to \mathbf{R}$, $i = 1, \ldots, d$, *such that* $w^i(t) - \int_0^t b^i(s)ds$, $t \geqq 0$, $i = 1, \ldots, d$, *are* $\mathscr{F}^\mu_t$-*martingales and Eqs. (2), (3), are satisfied. Then* $\mu \in \mathscr{R}(G, x_0)$.

Proof (1) For any $R > 0$ let

$$\tau_R(w) = \inf\{t \geqq 0; \; |w(t)| > R\}.$$

Then τ_R be an $\mathscr{F}^\mu_t$-stopping time. For any $f \in C^\infty(\mathbf{R}^d)$, let $X^f : [0, \infty) \times W^d \to \mathbf{R}$ be given by

$$X^f(t,w) = f(w(t)) - f(w(0)) - \int_0^t (L_G f)(s,w)ds, \qquad t \geqq 0,\ w \in W^d.$$

Then it is easy to see that $(X^f)^{\tau_R}$ is a supermartingale for any $R > 0$. In particular, X^f is a semi-martingale. Applying this for $f^i(x) = x^i$, $x \in \mathbf{R}$, we see that

$$w^i(t) - x_0^i = X^{f^i}(t) + \int_0^t (L_G f^i)(s,w)ds$$

is a semimartingale. Let M^i and A^i be a local martingale part and finite total variation part of $w^i - x_0^i$, $i = 1, \ldots, d$, respectively. Then by Ito's lemma, we see that

$$X^f(t,w) = \sum_{i=1}^d \int_0^t \frac{\partial f}{\partial x^i}(w(s))dM^i(s) + \sum_{i=1}^d \int_0^t \frac{\partial f}{\partial x^i}(w(s))dA^i(s)$$

$$+\frac{1}{2}\sum_{i,j=1}^d \int_0^t \frac{\partial^2 f}{\partial x^i \partial x^j}(w(s))d\langle M^i, M^j\rangle(s) - \int_0^t (L_G f)(s,w)ds,$$

and so

$$\sum_{i=1}^d \int_0^t \frac{\partial f}{\partial x^i}(w(s))dA^i(s) + \frac{1}{2}\sum_{i,j=1}^d \int_0^t \frac{\partial^2 f}{\partial x^i \partial x^j}(w(s))d\langle M^i, M^j\rangle(s)$$

$$- \int_0^t (L_G f)(s,w)ds$$

is a non-increasing process.

Applying this to $f = f^i$ and $f = -f^i$, $i = 1, \ldots, d$, we see that $A^i(t) - \int_0^t (L_G f^i)(s,w)ds$ and $-A^i(t) - \int_0^t (L_G(-f^i))(s,w)ds$ are non-increasing. So we see that $A^i(t)$ is absolutely continuous in t. Similarly applying this to $f(x) = x^i x^j$ and $f(x) = -x^i x^j$, we see that $\langle M^i, M^j\rangle(t)$ is absolutely continuous in t. So we see that there are $\{\mathscr{F}_t^\mu\}_{t\geqq 0}$-progressively measurable processs $a^{ij} : [0,\infty) \times W^d \to \mathbf{R}$, $i, j = 1, \ldots, d$, and $b^i : [0,\infty) \times W^d \to \mathbf{R}$, $i = 1, \ldots, d$, such that

$$A^i(t) = \int_0^t b^i(s)ds, \qquad t \geqq 0,\ i = 1, \ldots, d,$$

and

$$\langle M^i, M^j\rangle_t = \int_0^t a^{ij}(s)ds,\ i,j = 1, \ldots, d, \qquad t \geqq 0,\ \mu - a.s.w.$$

Note that for any $f \in C^\infty(\mathbf{R}^d)$

$$\int_0^t ((\frac{1}{2}(\nabla^2 f)(w(s)), (\nabla f)(w(s))), (\{a^{ij}(s)\}_{i,j=1,\ldots,d}, (b^i(s))_{i=1,\ldots d}))_{E^d} ds$$

$$- \int_0^t G(s, w, (\frac{1}{2}(\nabla^2 f)(w(s)), (\nabla f)(w(s))))ds$$

is a non-increasing process. Therefore

$$((\frac{1}{2}(\nabla^2 f)(w(t)), (\nabla f)(w(t))), (\{a^{ij}(t)\}_{i,j=1,\ldots,d}, (b^i(t))_{i=1,\ldots d}))_{E^d}$$

$$\leqq G(t, w, (\frac{1}{2}(\nabla^2 f)(w(t)), (\nabla f)(w(t)))) \ a.e.t \ \mu - a.s.w.$$

Let $\mathscr{H}$ be a countable subset of $C^\infty(\mathbf{R}^d)$ given by

$$\mathscr{H} = \{\sum_{i,j=1}^d q_{ij} x^i x^j + \sum_{i=1}^d r_i x^i; \ q_{ij}, r_i \in \mathbf{Q}, \ i, j = 1, \ldots, d\}.$$

Then we see that for any $x \in \mathbf{R}$, $\{(\frac{1}{2}(\nabla^2 f)(x), (\nabla f)(x)); \ f \in \mathscr{H}\}$ is dense in E^d. So we see that

$$(\{a^{ij}(t)\}_{i,j=1,\ldots,d}, (b^i(t))_{i=1,\ldots d}) \in K_G(t, w(t)) \quad a.e.t \ \mu - a.s.w$$

(2) From the assumption, we see that $K_G(t, w) \subset B_{E^d}(0, C)$. Then for any $f \in C_0^\infty(\mathbf{R}^d)$,

$$f(w(t)) - \int_0^t ((\frac{1}{2}(\nabla^2 f)(w(s)), (\nabla f)(w(s))), (\{a^{ij}(s)\}_{i,j=1,\ldots,d}, (b^i(s))_{i=1,\ldots d}))_{E^d} ds$$

is a martingale under μ. Since

$$((\frac{1}{2}(\nabla^2 f)(w(t)), (\nabla f)(w(t))), (\{a^{ij}(t)\}_{i,j=1,\ldots,d}, (b^i(t))_{i=1,\ldots d}))_{E^d}$$

$$\leqq G(t, w, (\frac{1}{2}(\nabla^2 f)(w(t)), (\nabla f)(w(t)))) \ a.e.t \ \mu - a.s.w$$

for any $f \in C_0^\infty(\mathbf{R}^N)$, we have Assertion (2). ■

Theorem 2 *Let $G \in \mathscr{H}\mathscr{J}^d$. Assume that $G : [0, \infty) \times W^d \times E^d \to \mathbf{R}$ is continuous and that there is a constant $C > 0$ such that $|G(t, w, z)| \leqq C|z|_{E^d}$ for any $(t, w, z) \in [0, \infty) \times W^d \times E^d$. Then we see that $\mathscr{R}(G, x) \neq \emptyset$ for any $x \in \mathbf{R}^d$.*

Proof Let us take an $e_0 \in E^d$. Let $a : [0, \infty) \times W^d \to S^d$ and $b : [0, \infty) \times W^d \to \mathbf{R}^d$ be given by

$$(a(t, w), b(t, w)) = P(e_0, K_G(t, w)), \qquad (t, w) \in [0, \infty) \times W^d.$$

Then by Proposition 16, we see that $a(t, w) \in S^d_+$ for all $(t, w) \in [0, \infty) \times W^d$. Also, by Propositions 4 and 6, we see that $a : [0, \infty) \times W^d \to S^d$ and $b : [0, \infty) \times W^d \to \mathbf{R}^d$ are bounded continuous functions. Let $\sigma : [0, \infty) \times W^d \to S^d$ be given by $\sigma(t, w) = a(t, w)^{1/2}$, $(t, w) \in [0, \infty) \times W^d$. Then by Proposition 15, we see that $\sigma : [0, \infty) \times W^d \to S^d$ is bounded and continuous. Then by Theorem 2.2 in Ikeda–Watanabe [2, p. 169], we see that for any $x \in \mathbf{R}^d$ there is a probability measure μ on W^d such that $\mu(w(0) = x) = 1$, $w^i(t) - \int_0^t b^i(s, w)ds$, $i = 1, \ldots, d$, are martingales under μ, and

$$\langle w^i, w^j \rangle_t = \int_0^t a^{ij}(s, w)ds \qquad \mu - a.s.w$$

So by Proposition 17, we see that $\mu \in \mathscr{R}(G, x)$. ■

5 Stability

Since the set $\mathscr{R}(G, x)$ contains many measures in general, we cannot discuss the uniqueness of a solution. However, in martingale problems the uniqueness of a solution and the stability of solutions are strongly related (c.f. Kaneko–Nakao [3]). So we discuss the stability of solutions.

Proposition 18 *Let $G_\infty, G_n \in \mathscr{H}\mathscr{J}^d$, $n = 1, 2, \ldots$, and assume the following.*

(1) *There is an $R < \infty$ such that*

$$|G_n(t, w, \xi)| \leqq R|\xi|_{E^d}, \qquad (t, w, \xi) \in [0, \infty) \times W^d \times E^d, \ n = 1, 2, \ldots, \infty.$$

(2) *For any $t \geqq 0$, $G_n(t, \cdot, \cdot) : W^d \times E^d \to \mathbf{R}$, $n = 1, 2, \ldots, \infty$, are continuous.*
(3) *For any $t \geqq 0$, $G_n(t, w, \xi) \to G_\infty(t, w, \xi)$, $n \to \infty$, uniformly in compacts with respect to $(w, \xi) \in W^d \times E^d$.*

Suppose moreover that $x_\infty, x_n \in \mathbf{R}^d$, $n = 1, 2, \ldots$, and $x_n \to x_\infty$, $n \to \infty$, and that $\mu_n \in \mathscr{R}(G_n, x_n)$, $n = 1, 2, \ldots$ Then there are a subsequence $\{n_k\}_{k=1}^\infty$ and $\mu_\infty \in \mathscr{R}(G_\infty, x_\infty)$ such that $\mu_{n_k} \to \mu_\infty$, $k \to \infty$.

Proof From the assumption and Proposition 2, we see that $K_{G_n}(t, w) \subset B_{E^d}(0, R)$ for any $n = 1, 2, \ldots, \infty$, $t \in [0, \infty)$ and $w \in W^d$. Then by Proposition 17 that $\bigcup_{n=1}^\infty \mathscr{R}(G_n, x_n)$ is relatively compact in $\mathscr{P}(W^d)$. So there are a subsequence $\{n_k\}_{k=1}^\infty$

and $\mu \in \mathscr{P}(W^d)$ such that $\mu_{n_k} \to \mu$ as $k \to \infty$. Therefore we see that for any $m \geqq 1$, $t > s \geqq s_m > \cdot > s_1 \geqq 0$, $f \in C_0^\infty(\mathbf{R}^d)$ and $g \in C_b(\mathbf{R}^{dm})$ with $g \geqq 0$,

$$E^{\mu}[(f(w(t)) - f(w(s)) - \int_s^t (L_{G_\infty})f)(r,w)dr)g(w(s_1),\cdots,w(s_m))]$$

$$= \lim_{k\to\infty} E^{\mu_{n_k}}[(f(w(t)) - f(w(s)) - \int_s^t (L_{G_{n_k}} f)(r,w)dr)g(w(s_1),\cdots,w(s_m))] \leqq 0.$$

This shows that $\mu \in \mathscr{R}(G_\infty, x_\infty)$. ■

For $d, r \geqq 1$, let $\mathscr{M}^{d\times r}$ be the set of $d \times r$ real matrices. Let $\hat{E}^{d,r} = \mathbf{R}^d \times \mathscr{M}^{d\times r}$ and we define innerproduct on $\hat{E}^{d,r}$ by

$$((b_1, C_1), (b_2, C_2))_{\hat{E}^{d,r}} = (b_1, b_2)_{\mathbf{R}^d} + trace(C_1 C_2^*)$$

for $(b_1, C_1), (b_2, C_2) \in \mathbf{R}^d \times \mathscr{M}^{d\times r}$. Also we define $q_{d,r} : \hat{E}^{d,r} \to E^d$ by

$$q_{d,r}((b, C)) = (CC^*, b), \qquad (b, C) \in \hat{E}^{d,r}.$$

Then we have the following.

Theorem 3 *Let $G_\infty, G_n \in \mathscr{H}\mathscr{J}^d$, $n = 1, 2, \ldots, x_\infty, x_n \in \mathbf{R}^d$, $n = 1, 2, \ldots$, and assume the following.*

(1) *There is a $C_0 < \infty$ such that*

$$|G_n(t, w, \xi)| \leqq C_0|\xi|_{E^d}, \qquad (t, w, \xi) \in [0, \infty) \times W^d \times E^d, \; n = 1, 2, \ldots, \infty.$$

(2) *For any $t \geqq 0$, $G_n(t, \cdot, \cdot) : W^d \times E^d \to \mathbf{R}$, $n = 1, 2, \ldots, \infty$ are continuous.*
(3) *There are $r \geqq 1$, and measurable maps $V_n : [0, \infty) \times W^d \times E^d \to \hat{E}^{d,r}$, $n = 1, 2, \ldots, \infty$ satisfying the following.*

(i) *There is a $C_1 < \infty$ such that for any $t \in [0, \infty)$, $w, w' \in W^d$ and $\xi \in E^d$,*

$$|V_n(t, w, \xi) - V_n(t, w', \xi)|_{\hat{E}^{d,r}} \leqq C_1 \sup\{|w(s) - w'(s)|;\; s \in [0, t]\}$$

for $n = 1, 2, \ldots, \infty$.
(ii) *For any $n = 1, 2, \ldots, \infty$, $t \in [0, \infty)$, $w \in W^d$ and $\xi \in E^d$,*

$$q_{d,r}(V_n(t, w, \xi)) \in K_{G_n}(t, w)$$

and

$$q_{d,r}(V_n(t, w, \xi)) = \xi, \;\; if\, \xi \in K_{G_n}(t, w).$$

(iii) *For any* $t \in [0,\infty)$ *and* $(w,\xi) \in W^d \times E^d$, $V_n(t,w,\xi) \to V_\infty(t,w,\xi)$, $n \to \infty$. *Also,* $x_n \to x_\infty$, $n \to \infty$.

(4) *For any* $t \geqq 0$, $G_n(t,w,\xi) \to G_\infty(t,w,\xi)$, $n \to \infty$, *uniformly in compacts with respect to* $(w,\xi) \in W^d \times E^d$.

Then $\mathscr{R}(G_n, x_n) \to \mathscr{R}(G_\infty, x_\infty)$ *as compact sets in* $\mathscr{P}(W^d)$ *with respect to the Hausdorff metrics.*

Proof Since we already have Proposition 18, it is sufficient to show that for any $\mu \in \mathscr{R}(G_\infty, x_\infty)$ there are $\mu_n \in \mathscr{R}(G_n, x_n)$ such that $\mu_n \to \mu$, $n \to \infty$, in $\mathscr{P}(W^d)$. So let $\mu \in \mathscr{R}(G_\infty, x_\infty)$.

Then we see by Propositions 17 that there are $\{\mathscr{F}_t^\mu\}_{t\geqq 0}$-progressively measurable processs $\tilde{b}^i : [0,\infty) \times W^d \to \mathbf{R}$, $i = 1,\ldots,d$, and $\tilde{a}^{ij} : [0,\infty) \times W^d \to \mathbf{R}$, $i,j = 1,\ldots,d$, such that $w^i(t) - \int_0^t \tilde{b}^i(s)ds$, $t \geqq 0$, $i = 1,\ldots,d$, are $\mathscr{F}_t^\mu$-martingales,

$$\langle w^i, w^j \rangle_t = \int_0^t \tilde{a}^{ij}(s)ds,\ i,j = 1,\ldots,d, \qquad t \geqq 0,\ \mu - a.s.w,$$

and

$$(\{\tilde{a}^{ij}(t)\}_{i,j=1,\ldots,d}), (\tilde{b}^i(t))_{i=1,\ldots,d})) \in K_{G_\infty}(t,w) \qquad a.e.t \geqq 0,\ \mu - a.s.w.$$

Let $(\hat{b}(t), \hat{\sigma}(t)) = V_\infty(t, w, (\tilde{b}(t), \tilde{a}(t)))$. Then from the assumption (3)(ii) we see that $\tilde{b}(t) = \hat{b}(t)$ and $\tilde{a}(t) = \hat{\sigma}(t)\hat{\sigma}(t)^*$.

Then by Representation Theorem (e.g. Theorem 7.1′ in Ikeda–Watanabe [2] p. 90) we see that there are a filtered probability space $(\Omega, \mathscr{F}, P, \{\mathscr{F}_t\}_{t\in[0,\infty)})$ with usual condition, an r-dimensional $\{\mathscr{F}_t\}$-Brownian motion $\{B_t\}_{t\in[0,\infty)}$, a continuous adapted process $X : [0,\infty) \times \Omega \to \mathbf{R}^d$, progressively measurable processs $b : [0,\infty) \times \Omega \to \mathbf{R}^d$ and $\sigma : [0,\infty) \times \Omega \to \mathscr{M}^{d\times r}$ such that

$$X^i(t) = x_\infty^i + \int_0^t b^i(s)ds + \sum_{k=1}^r \int_0^t \sigma^{ik}(s)dB^k(s), \qquad i = 1,\ldots,d,\ t \in [0,\infty)$$

$$(b(t), \sigma(t)) = V_\infty(t, X(\cdot), (\tilde{b}(t, X(\cdot)), \tilde{a}(t, X(\cdot))),$$

$$q_{d,r}(b(t, X(\cdot)), \sigma(t, X(\cdot))) \in K_{G_\infty}(t, X(\cdot))\ a.s.\omega\ a.e.t$$

and that the probability law of $X(\cdot)$ is μ.

Let $z : [0,\infty) \times \Omega \to E^d$ be given by $z(t) = (\tilde{b}(t, X(\cdot)), \tilde{a}(t, X(\cdot)))$, $t \in [0,\infty)$. Then we see that X is a solution to the following SDE

$$X(t) = x_\infty + \int_0^t V_\infty(s, X(\cdot), z(s))dB(s), \qquad s \in [0,\infty).$$

Here we use the notation that for any progressively measurable processes $\eta = (\bar{b}, \bar{\sigma})$: $[0, \infty) \times \Omega \to \mathbf{R}^d \times \mathscr{M}^{d \times r}$

$$\int_0^t \eta(s)dB(s) = \int_0^t \bar{b}(s)ds + \sum_{k=1}^r \int_0^t \bar{\sigma}_k(s)dB^k(s).$$

Now let us think of the following SDE for each $n = 1, 2, \ldots$

$$X_n(t) = x_n + \int_0^t V_n(s, X_n(\cdot), z(s))dB(s), \qquad s \in [0, \infty).$$

Since

$$|V_n(t, w, z(t))|^2_{\hat{E}^{d,r}} \leqq |q_{d,r}(V_n(t, w, z(t)))|_{E^d} + |q_{d,r}(V_n(t, w, z(t)))|^2_{E^d} \leqq C_0^{1/2} + C_0$$

and

$$|V_n(t, w, z(t)) - V_n(t, w', z(t))|_{\hat{E}^{d,r}} \leqq C_1 \max_{s \in [0,t]} |w(s) - w'(s)|$$

for any $n = 1, 2, \ldots, t \in [0, \infty)$ and $w, w' \in W^d$, we see that there is a path-wise unique solution X_n. Let $\mu_n \in \mathscr{P}(W^d)$ be the probability law of $X_n(\cdot)$. Since

$$q_{d,r}(V_n(s, X_n(\cdot), z(s))) \in K_{G_n}(t, X_n(\cdot)),$$

we see by Proposition 17 that $\mu_n \in \mathscr{R}(G_n, x_n)$, $n = 1, 2, \ldots$ Also we have

$$\begin{aligned} &|X(t) - X_n(t)| \\ &\leqq |x_\infty - x_n| + |\int_0^t (V_\infty(s, X(\cdot), z(s)) - V_n(s, X(\cdot), z(s))dB(s)| \\ &\qquad + |\int_0^t (V_n(s, X(\cdot), z(s)) - V_n(s, X_n(\cdot), z(s))dB(s)|. \end{aligned}$$

Therefore we see that

$$\begin{aligned} &E[\sup_{s \in [0,t]} |X(s) - X_n(s)|^2] \\ &\leqq r_n(T) + 3(4d^2 + Td) \int_0^t E[|V_n(s, X(s), z(s)) - V_n(s, X_n(s), z(s))|^2_{\hat{E}^{d,r}}]ds \\ &\leqq r_n(T) + 3(4d^2 + Td)C_1^2 \int_0^t E[\sup_{s \in [0,r]} |X(s) - X_n(s)|^2]dr \end{aligned}$$

for any $t \in [0, T]$, $T > 0$, where

$$r_n(T) = 3|x_\infty - x_n|^2 + 3(4d^2 + Td) \int_0^T E[|V_\infty(s, X(\cdot), z(s)) - V_n(s, X(\cdot), z(s))|^2_{\hat{E}^{d,r}}]ds.$$

Since we see from the assumption that $r_n(T) \to 0$, $n \to \infty$, we see by Gronwall's inequality that

$$E[\sup_{s\in[0,t]} |X(s) - X_n(s)|^2] \to 0, \ n \to \infty, \qquad t \in [0, T], \ T > 0.$$

This implies that $\mu_n \to \mu$, $n \to \infty$, in $\mathscr{P}(W^d)$. So we have our assertion. ■

For any $\gamma > 0$, let $S^d_{\gamma+}$ be the set of $A \in S^d_+$ such that the minimum eigenvalue of $\lambda_{min}(A)$ of A is greater than or equal to γ.

Then we have the following.

Corollary 1 *Let* $G_\infty, G_n \in \mathscr{H}\mathscr{J}^d$, $n = 1, 2, \ldots, x_\infty, x_n \in \mathbf{R}^d$, $n = 1, 2, \ldots$, *and assume the following.*

(1) *For any* $t \geqq 0$, $G_n(t, \cdot, \cdot) : W^d \times E^d \to \mathbf{R}$, $n = 1, 2, \ldots, \infty$, *are continuous.*
(2) *There is a* $C_0 \in (0, \infty)$ *such that*

$$|G_n(t, w, \xi)| \leqq C_0|\xi|_{E^d},$$

and

$$|G_n(t, w, \xi) - G_n(t, w', \xi)| \leqq C_0|\xi|_{E^d} \max_{s\in[0,t]} |w(s) - w'(s)|$$

for any $t \in [0, \infty)$, $w, w' \in W^d$, $\xi \in E^d$, *and* $n = 1, 2, \ldots, \infty$.
(3) *There are* $R_1 > R_0 > 0$ *and* $\gamma > 0$ *such that* $K_{G_n}(t, w) \in \mathscr{K}_{R_0,R_1}(E^d)$ *and* $K_{G_n}(t, w) \subset S^d_{\gamma+} \times \mathbf{R}^d$ *for any* $n = 1, 2, \ldots, \infty$, $t \in [0, \infty)$ *and* $w \in W^d$.
(4) *For any* $t \geqq 0$, $G_n(t, w, \xi) \to G_\infty(t, w, \xi)$, $n \to \infty$, *uniformly in compacts with respect to* $(w, \xi) \in W^d \times E^d$. *Also,* $x_n \to x_\infty$, $n \to \infty$.

Then $\mathscr{R}(G_n, x_n) \to \mathscr{R}(G_\infty, x_\infty)$ *as compact sets in* $\mathscr{P}(W^d)$ *with respect to the Hausdorff metrics.*

Proof Let us define $\psi_d : E^d \to \mathbf{R}^d \times \mathscr{M}^{d\times d}$ by

$$\psi_d(A, b) = (b, A^{1/2}), \qquad (A, b) \in E^d.$$

Now let us define $V_n : [0, \infty) \times W^d \times E^d \to \mathbf{R}^d \times \mathscr{M}^{d\times d}$, $n = 1, 2, \ldots, \infty$, by

$$V_n(t, w, z) = \psi_d(\Psi(z, K_{G_n}(t, w))), \qquad t \in [0, \infty), \ w \in W^d, \ z \in E^d.$$

Then by Propositions 4 and 15(3), and Theorem 1, we see that V_n's satisfy the assumption (3) in Theorem 3. This implies our assertion. ■

References

1. Boyle PP, Vorst T (1992) Option replication in discrete time with transaction costs. J Financ 47:271–293
2. Ikeda N, Watanabe S (1989) Stochastic differential equations and difusion processes, 2nd edn. Kodansha North-Holland
3. Kaneko H, Nakao S (1988) A note on approximation for stochastic differential equations. Séminare de Probabilités, XXII, Lecture notes in mathematics, vol 1321. Springer, Berlin, pp 155–162
4. Kusuoka S (1995) Limit theorem on option replication cost with transaction costs. Ann Appl Prob 5:198–221

Bolza Optimal Control Problems with Linear Equations and Periodic Convex Integrands on Large Intervals

Alexander J. Zaslavski

Abstract We study the structure of approximate solutions of Bolza optimal control problems, governed by linear equations, with periodic convex integrands, on large intervals, and show that the turnpike property holds. To have this property means, roughly speaking, that the approximate optimal trajectories are determined mainly by the integrand, and are essentially independent of the choice of time intervals and data, except in regions close to the endpoints of the time interval. We also show the stability of the turnpike phenomenon under small perturbations of integrands and study the structure of approximate optimal trajectories in regions close to the endpoints of the time intervals.

Keywords Good trajectory-control pair · Integrand · Optimal control problem · Overtaking optimal trajectory-control pair · Turnpike property

Article type: Research Article
Received: September 6, 2016
Revised: January 9, 2017

1 Introduction

The growing significance of the study of (approximate) solutions of variational and optimal control problems defined on infinite intervals and on sufficiently large intervals has been realized in the recent years [2, 4–11, 15, 18, 19, 21, 22, 25–27]. This is due not only to theoretical achievements in this area, but also because of

---

JEL Classification: C02, C61, C67.
Mathematics Subject Classification (2010): 49J15, 49J99, 90C26, 90C31, 93C15.

A.J. Zaslavski (✉)
Department of Mathematics, The Technion – Israel Institute of Technology,
Technion City, 32000 Haifa, Israel
e-mail: ajzasl@techunix.technion.ac.il

S. Kusuoka and T. Maruyama (eds.), *Advances in Mathematical Economics*, Advances in Mathematical Economics 21,
DOI 10.1007/978-981-10-4145-7_4

numerous applications to engineering [1, 6, 22, 28], models of economic dynamics [6, 14, 17, 20, 22, 25, 27] the game theory [12, 22, 24, 25], models of solid-state physics [3] and the theory of thermodynamical equilibrium for materials [13, 16]. In [26], for the Lagrange optimal control problems, governed by linear equations, with nonautonomous periodic convex integrands, on large intervals we proved that the turnpike phenomenon holds and described the structure of approximate optimal trajectories in regions close to the endpoints of the time intervals. It was established that in these regions optimal trajectories converge to solutions of the corresponding infinite horizon optimal control problem which depend only on the integrand. In the present paper we generalize these results for Bolza problems.

We study the structure of approximate optimal trajectories of linear control systems

$$x'(t) = Ax(t) + Bu(t), \tag{1.1}$$

$$x(0) = x_0$$

with periodic convex integrands $f : [0, \infty) \times R^n \times R^m \to R^1$, where A and B are given matrices of dimensions $n \times n$ and $n \times m, x(t) \in R^n, u(t) \in R^m$ and the admissible controls are Lebesgue measurable functions.

We assume that the linear system (1.1) is controllable and that the integrand f is a Borel measurable function.

We denote by $|\cdot|$ the Euclidean norm and by $\langle\cdot,\cdot\rangle$ the inner product in the n-dimensional Euclidean space R^n. Denote by $\mathbf{Z}$ the set of all integers. For every $z \in R^1$ denote by $\lfloor z \rfloor$ the largest integer which does not exceed z: $\lfloor z \rfloor = \max\{i \in \mathbf{Z} : i \le z\}$.

The performance of the above control system is measured on any finite interval $[T_1, T_2] \subset [0, \infty)$ by the integral functional

$$I^f(T_1, T_2, x, u) = \int_{T_1}^{T_2} f(t, x(t), u(t))dt. \tag{1.2}$$

We suppose that the integrand $f : [0, \infty) \times R^n \times R^m \to R^1$ satisfies the following Assumption (A)

(i) $f(t + \tau, x, u) = f(t, x, u)$ for all $t \in [0, \infty)$, all $x \in R^n$ and all $u \in R^m$ for some constant $\tau > 0$ depending only on f;
(ii) for any $t \in [0, \infty)$ the function $f(t, \cdot, \cdot) : R^n \times R^m \to R^1$ is strictly convex;
(iii) the function f is bounded on any bounded subset of $[0, \infty) \times R^n \times R^m$;
(iv) $f(t, x, u) \to \infty$ as $|x| \to \infty$ uniformly in $(t, u) \in [0, \infty) \times R^m$;
(v) $f(t, x, u)|u|^{-1} \to \infty$ as $|u| \to \infty$ uniformly in $(t, x) \in [0, \infty) \times R^n$.

Assumption (A) implies that f is bounded below on $[0, \infty) \times R^n \times R^m$.

Let $T_2 > T_1 \ge 0$. A pair of an absolutely continuous (a.c.) function $x : [T_1, T_2] \to R^n$ and a Lebesgue measurable function $u : [T_1, T_2] \to R^m$ is called an (A, B)-trajectory-control pair if for almost every (a. e.) $t \in [T_1, T_2]$ (1.1) holds.

Denote by $X(A, B, T_1, T_2)$ the set of all (A, B)-trajectory-control pairs $x : [T_1, T_2] \to R^n, u : [T_1, T_2] \to R^m$.

Let $J = [a, \infty)$ be an infinite closed subinterval of $[0, \infty)$. A pair of functions $x : J \to R^n$ and $u : J \to R^m$ is called an (A, B)-trajectory-control pair if it is an (A, B)-trajectory-control pair on any bounded closed subinterval of J. Denote by $X(A, B, a, \infty)$ the set of all (A, B)-trajectory-control pairs $x : J \to R^n$, $u : J \to R^m$.

In Chap. 2 of [26] we study the structure of approximate optimal trajectories of the linear control system (1.1) with the integrand f and show that the turnpike property holds. To have this property means, roughly speaking, that the approximate optimal trajectories on sufficiently large intervals are determined mainly by the integrand, and are essentially independent of the choice of time intervals and data, except in regions close to the endpoints of the time intervals. In Chap. 2 of [26] we also show the stability of the turnpike phenomenon under small perturbations of the integrand and study the structure of approximate optimal trajectories in regions close to the endpoints of the time intervals.

More precisely, in Chap. 2 of [26] we consider the following optimal control problems

$$I^f(0, T, x, u) \to \min, \tag{P_1}$$

$$(x, u) \in X(A, B, 0, T) \text{ such that } x(0) = y, \; x(T) = z,$$

$$I^f(0, T, x, u) \to \min, \tag{P_2}$$

$$(x, u) \in X(A, B, 0, T) \text{ such that } x(0) = y,$$

$$I^f(0, T, x, u) \to \min, \tag{P_3}$$

$$(x, u) \in X(A, B, 0, T),$$

where $y, z \in R^n$ and $T > 0$. The study of these problems is based on the properties of solutions of the corresponding infinite horizon optimal control problem associated with the control system (1.1) and the integrand f.

In [28] we were interested in a turnpike property of the approximate solutions of problems (P_2). In Chap. 2 of [26] we established the turnpike property of the approximate solutions of problems (P_1) and (P_3), showed the stability of the turnpike phenomenon under small perturbations of the integrand f and studied the structure of approximate optimal trajectories in regions close to the endpoints of the time intervals.

For the problems (P_2) and (P_3) we showed that in regions close to the right endpoint T of the time interval these approximate solutions are determined only by the integrand, and are essentially independent of the choice of the interval and the endpoint value y. For the problems (P_3), approximate solutions are determined only by the integrand also in regions close to the left endpoint 0 of the time interval.

The following result was obtained in [28] (see also Chap. 6 of [22] and Chap. 2 of [26]).

Proposition 1 *There exists* $(x_f, u_f) \in X(A, B, 0, \tau)$ *which is the unique solution of the following minimization problem*

$$I^f(0, \tau, x, u) \to \min, \ (x, u) \in X(A, B, 0, \tau) \text{ such that } x(0) = x(\tau).$$

Let a trajectory-control pair $(x_f, u_f) \in X(A, B, 0, \tau)$ be as guaranteed by Proposition 1. Put

$$\mu(f) = \tau^{-1} I^f(0, \tau, x_f, u_f). \tag{1.3}$$

The following results were obtained in [28] (see also Chap. 6 of [22] and Chap. 2 of [26]).

Theorem 1 *For any* $(x, u) \in X(A, B, 0, \infty)$ *either*

$$\text{(i) } I^f(0, T, x, u) - T\mu(f) \to \infty \text{ as } T \to \infty$$

$$\text{or (ii) } \sup\{|I^f(0, T, x, u) - T\mu(f)| : T > 0\} < \infty.$$

Moreover, if relation (ii) holds, then

$$\sup\{|x(i\tau + t) - x_f(t)| : t \in [0, \tau]\} \to 0 \text{ as } i \to \infty, \text{ where } i \in \mathbf{Z}.$$

We say that $(x, u) \in X(A, B, 0, \infty)$ is (f, A, B)-good [22, 25, 26] if

$$\sup\{|I^f(0, T, x, u) - T\mu(f)| : T > 0\} < \infty.$$

The second statement of Theorem 1 describes the asymptotic behavior of (f, A, B)-good trajectory-control pairs, shows that the corresponding infinite horizon optimal control problem has the turnpike property and that the function x_f is its turnpike.

We say that $(\tilde{x}, \tilde{u}) \in X(A, B, 0, \infty)$ is (f, A, B)-overtaking optimal [22, 25, 26] if for each $(x, u) \in X(A, B, 0, \infty)$ satisfying $x(0) = \tilde{x}(0)$,

$$\limsup_{T\to\infty} [I^f(0, T, \tilde{x}, \tilde{u}) - I^f(0, T, x, u)] \le 0.$$

Theorem 2 *Let* $x_0 \in R^n$. *Then there is an* (f, A, B)*-overtaking optimal trajectory-control pair* $(\tilde{x}, \tilde{u}) \in X(A, B, 0, \infty)$ *satisfying* $\tilde{x}(0) = x_0$. *Moreover, if* $(x, u) \in X(A, B, 0, \infty) \setminus \{(\tilde{x}, \tilde{u})\}$ *satisfies* $x(0) = x_0$, *then there are* $T_0 > 0$ *and* $\varepsilon > 0$ *such that*

$$I^f(0, T, x, u) \ge I^f(0, T, \tilde{x}, \tilde{u}) + \varepsilon \text{ for all } T \ge T_0.$$

The next result describes the limit behavior of overtaking optimal trajectories.

Theorem 3 *Let $M, \varepsilon > 0$. Then there exists a natural number N such that for any (f, A, B)-overtaking optimal trajectory-control pair $(x, u) \in X(A, B, 0, \infty)$ which satisfies $|x(0)| \leq M$ the relation*

$$\sup\{|x(i\tau + t) - x_f(t)| : t \in [0, \tau]\} \leq \varepsilon \tag{1.4}$$

holds for all integers $i \geq N$. Moreover, there exists $\delta > 0$ such that for any (f, A, B)-overtaking optimal trajectory-control pair $(x, u) \in X(A, B, 0, \infty)$ satisfying $|x(0) - x_f(0)| \leq \delta$, the relation (1.4) holds for all integers $i \geq 0$.

Let $T > 0$ and $y, z \in R^n$. Set

$$\sigma(f, y, z, T) = \inf\{I^f(0, T, x, u) :$$

$$(x, u) \in X(A, B, 0, T) \text{ and } x(0) = y, \ x(T) = z\}, \tag{1.5}$$

$$\sigma(f, y, T) = \inf\{I^f(0, T, x, u) : \ (x, u) \in X(A, B, 0, T) \text{ and } x(0) = y\}, \tag{1.6}$$

$$\widehat{\sigma}(f, z, T) = \inf\{I^f(0, T, x, u) : \ (x, u) \in X(A, B, 0, T) \text{ and } x(T) = z\}, \tag{1.7}$$

$$\sigma(f, T) = \inf\{I^f(0, T, x, u) : \ (x, u) \in X(A, B, 0, T)\}. \tag{1.8}$$

It follows from assumption (A) and Proposition 2.28 of [26] that

$$-\infty < \sigma(f, y, z, T), \sigma(f, y, T), \widehat{\sigma}(f, z, T), \sigma(f, T) < \infty.$$

The next theorem establishes the turnpike property for approximate solutions of problems (P_2) with the turnpike $x_f(\cdot)$.

Theorem 4 *Let $M, \varepsilon > 0$. Then there exist an integer $N \geq 1$ and $\delta > 0$ such that for each $T > 2N\tau$ and each $(x, u) \in X(A, B, 0, T)$ which satisfies*

$$|x(0)| \leq M, \ I^f(0, T, x, u) \leq \sigma(f, x(0), T) + \delta$$

the inequality

$$\sup\{|x(i\tau + t) - x_f(t)| : t \in [0, \tau]\} \leq \varepsilon \tag{1.9}$$

holds for all integers $i \in [N, \tau^{-1}T - N]$. Moreover if $|x(0) - x_f(0)| \leq \delta$, then inequality (1.9) holds for all integers $i \in [0, \tau^{-1}T - N]$.

Theorems 1–4 were obtained in [28] (see also Chap. 6 of [22]). Note that under assumptions of Theorem 4, if $|x(\lfloor \tau^{-1}T \rfloor \tau) - x_f(0)| \leq \delta$, then inequality (1.9) holds for all integers $i \in [N, \tau^{-1}T - 1]$.

The next two results obtained in Chap. 2 of [26] establish the turnpike property for approximate solutions of problems (P_1) and (P_3) respectively with the turnpike $x_f(\cdot)$.

Theorem 5 *Let $M, \varepsilon > 0$. Then there exist an integer $N \geq 1$ and $\delta > 0$ such that for each $T > 2N\tau$ and each $(x, u) \in X(A, B, 0, T)$ which satisfies*

$$|x(0)|,\ |x(T)| \leq M,\ I^f(0, T, x, u) \leq \sigma(f, x(0), x(T), T) + \delta$$

inequality (1.9) holds for all integers $i \in [N, \tau^{-1}T - N]$. Moreover if $|x(0) - x_f(0)| \leq \delta$, then inequality (1.9) holds for all integers $i \in [0, \tau^{-1}T - N]$ and if $|x(\lfloor \tau^{-1}T \rfloor \tau) - x_f(0)| \leq \delta$, then inequality (1.9) holds for all integers $i \in [N, \tau^{-1}T - 1]$.

Theorem 6 *Let $\varepsilon > 0$. Then there exist an integer $N \geq 1$ and $\delta > 0$ such that for each $T > 2N\tau$ and each $(x, u) \in X(A, B, 0, T)$ which satisfies*

$$I^f(0, T, x, u) \leq \sigma(f, T) + \delta$$

inequality (1.9) holds for all integers $i \in [N, \tau^{-1}T - N]$. Moreover if $|x(0) - x_f(0)| \leq \delta$, then inequality (1.9) holds for all integers $i \in [0, \tau^{-1}T - N]$ and if $|x(\lfloor \tau^{-1}T \rfloor \tau) - x_f(0)| \leq \delta$, then inequality (1.9) holds for all integers $i \in [N, \tau^{-1}T - 1]$.

Theorems 4–6 are partial cases of Theorem 2.13 of [26]. The next theorem establishes a weak version of the turnpike property for approximate solutions of problems (P_1), (P_2) and (P_3) with the turnpike $x_f(\cdot)$.

Theorem 7 *Let $\varepsilon, M_0, M_1 > 0$. Then there exist natural numbers Q, l such that for each $T > Ql\tau$ and each $(x, u) \in X(A, B, 0, T)$ which satisfies at least one of the following conditions below*

$|x(0)|,\ |x(T)| \leq M_0,\ I^f(0, T, x, u) \leq \sigma(f, x(0), x(T), T) + M_1$;

$|x(0)| \leq M_0,\ I^f(0, T, x, u) \leq \sigma(f, x(0), T) + M_1$;

$I^f(0, T, x, u) \leq \sigma(f, T) + M_1$

there exist strictly increasing sequences of nonnegative integers

$$\{a_i\}_{i=1}^q,\ \{b_i\}_{i=1}^q \subset [0, \tau^{-1}T]$$

such that $q \leq Q$,

$$0 \leq b_i - a_i \leq l \text{ for all } i = 1, \dots, q,$$

$b_i \leq a_{i+1}$ for all integers i satisfying $1 \leq i < q$ and that for each integer $i \in [0, \tau^{-1}T - 1] \setminus \cup_{j=1}^q [a_j, b_j]$,

$$|x(i\tau + t) - x_f(t)| \leq \varepsilon,\ t \in [0, \tau].$$

Theorem 7 is a partial case of Theorem 2.14 of [26], a stability result.

We say that $(x, u) \in X(A, B, 0, \infty)$ is (f, A, B)-minimal [3, 25, 26] if for each $T > 0$,

$$I^f(0, T, x, u) = \sigma(f, x(0), x(T), T). \tag{1.10}$$

The next result which is proved in Sect. 2.5 of [26] shows the equivalence of the optimality criterions introduced above.

Theorem 8 *Assume that* $(x, u) \in X(A, B, 0, \infty)$. *Then the following conditions are equivalent:*

(i) (x, u) *is* (f, A, B)*-overtaking optimal; (ii)* (x, u) *is* (f, A, B)*-minimal and* (f, A, B)*-good; (iii)* (x, u) *is* (f, A, B)*-minimal and*

$$\max\{|x(i\tau + t) - x_f(t)| : t \in [0, \tau]\} \to 0 \text{ as integers } i \to \infty;$$

(iv) (x, u) *is* (f, A, B)*-minimal and* $\liminf_{t\to\infty} |x(t)| < \infty$.

The following result is also proved in Sect. 2.5 of [26]. It shows that if the integrand f does not depend on the variable t, then $x_f(\cdot)$ is a constant function.

Theorem 9 *Assume that for each* $x \in R^n$, *each* $u \in R^m$ *and each* $t_1, t_2 \geq 0$, $f(t_1, x, u) = f(t_2, x, u)$. *Then* $x_f(t) = x_f(0)$ *for all* $t \in [0, \tau]$ *and* $x_f(0)$ *does not depend of* τ.

Corollary 1 *Assume that for each* $x \in R^n$, *each* $u \in R^m$ *and each* $t_1, t_2 \geq 0$, $f(t_1, x, u) = f(t_2, x, u)$. *Then for all* $t \in [0, \tau]$, $x_f(t) = x_*$ *and* $u_f(t) = u_*$ *where* $(x_*, u_*) \in R^n \times R^m$ *is a unique solution of the minimization problem*

$$f(x, u) \to \min, \ (x, u) \in R^n \times R^m, \ Ax + Bu = 0.$$

Note that autonomous integrands are used in order to determine an objective function in models of economic growth with a technology which does not depend on time. Nonautonomous periodic integrand can be used for models of economic growth with a time-depending technology under corresponding periodicity assumptions. This periodicity can occur if one take into account that every technology has several steps of developments and usage.

2 Stability of the Turnpike Phenomenon

In this section we state Theorems 2.12–2.14 of [26] which show that the turnpike phenomenon is stable under small perturbations of the integrand f. We use the notation, definitions and assumptions introduced in Sect. 1.

Recall that $f : [0, \infty) \times R^n \times R^m \to R^1$ is a Borel measurable function satisfying assumption (A). Let $a > 0$ and $\psi : [0, \infty) \to [0, \infty)$ be an increasing function such that

$$\lim_{t\to\infty} \psi(t) = \infty. \tag{2.1}$$

We suppose that for all $(t, x, u) \in [0, \infty) \times R^n \times R^m$,

$$f(t, x, u) \geq \max\{\psi(|x|),\ \psi(|u|)|u|\} - a. \tag{2.2}$$

Denote by $\mathcal{M}$ the set of all Borel measurable functions $g : [0, \infty) \times R^n \times R^m \to R^1$ which are bounded on all bounded subsets of $[0, \infty) \times R^n \times R^m$ and such that for all $(t, x, u) \in [0, \infty) \times R^n \times R^m$,

$$g(t, x, u) \geq \max\{\psi(|x|),\ \psi(|u|)|u|\} - a. \tag{2.3}$$

For the set $\mathcal{M}$ we consider the uniformity which is determined by the following base:

$$E(N, \varepsilon, \lambda) = \{(g_1, g_2) \in \mathcal{M} \times \mathcal{M} :\ |g_1(t, x, u) - g_2(t, x, u)| \leq \varepsilon \text{ for each } t \geq 0,$$

$$\text{each } x \in R^n \text{ satisfying } |x| \leq N \text{ and each } u \in R^m \text{ satisfying } |u| \leq N\}$$

$$\cap\{(g_1, g_2) \in \mathcal{M} \times \mathcal{M} :\ (|g_1(t, x, u)| + 1)(|g_2(t, x, u)| + 1)^{-1} \in [\lambda^{-1}, \lambda]$$

$$\text{for each } t \geq 0,\ \text{each } x \in R^n \text{ satisfying } |x| \leq N \text{ and each } u \in R^m\}, \tag{2.4}$$

where $N > 0$, $\varepsilon > 0$, $\lambda > 1$. It is not difficult to see that the space $\mathcal{M}$ with this uniformity is metrizable and complete.

Let $T_2 > T_1 \geq 0$, $y, z \in R^n$ and $g \in \mathcal{M}$. For each pair of Lebesgue measurable functions $x : [T_1, T_2] \to R^n$, $u : [T_1, T_2] \to R^m$ set

$$I^g(T_1, T_2, x, u) = \int_{T_1}^{T_2} g(t, x(t), u(t))dt \tag{2.5}$$

and set

$$\sigma(g, y, z, T_1, T_2) = \inf\{I^g(T_1, T_2, x, u) :$$

$$(x, u) \in X(A, B, T_1, T_2) \text{ and } x(T_1) = y,\ x(T_2) = z\}, \tag{2.6}$$

$$\sigma(g, y, T_1, T_2) = \inf\{I^g(T_1, T_2, x, u) :$$

$$(x, u) \in X(A, B, T_1, T_2) \text{ and } x(T_1) = y\}, \tag{2.7}$$

$$\widehat{\sigma}(g, z, T_1, T_2) = \inf\{I^g(T_1, T_2, x, u) :$$

$$(x, u) \in X(A, B, T_1, T_2) \text{ and } x(T_2) = z\}, \tag{2.8}$$

$$\sigma(g, T_1, T_2) = \inf\{I^g(T_1, T_2, x, u) :\ (x, u) \in X(A, B, T_1, T_2)\}. \tag{2.9}$$

Since any $g \in \mathcal{M}$ is bounded on all the bounded subsets of $[0, \infty) \times R^n \times R^m$ it follows from Proposition 2.28 of [26] and (2.3) that all the values defined above are finite.

In Chap. 2 of [26] we proved the following three stability results.

Theorem 10 *Let $\varepsilon, M > 0$. Then there exist an integer $L_0 \geq 1$ and $\delta_0 > 0$ such that for each integer $L_1 \geq L_0$ there exists a neighborhood $\mathcal{U}$ of f in $\mathcal{M}$ such that the following assertion holds.*

Assume that $T > 2L_1\tau$, $g \in \mathcal{U}$, $(x, u) \in X(A, B, 0, T)$ and that a finite sequence of integers $\{S_i\}_{i=0}^q$ satisfy

$$S_0 = 0,\ S_{i+1} - S_i \in [L_0, L_1],\ i = 0, \dots, q-1,\ S_q\tau \in (T - L_1\tau, T], \tag{2.10}$$

$$I^g(S_i\tau, S_{i+1}\tau, x, u) \leq (S_{i+1} - S_i)\tau\mu(f) + M$$

for each integer $i \in [0, q-1]$,

$$I^g(S_i\tau, S_{i+2}\tau, x, u) \leq \sigma(g, x(S_i\tau), x(S_{i+2}\tau), S_i\tau, S_{i+2}\tau) + \delta_0$$

for each nonnegative integer $i \leq q-2$ and

$$I^g(S_{q-2}\tau, T, x, u) \leq \sigma(g, x(S_{q-2}\tau), x(T), S_{q-2}\tau, T) + \delta_0.$$

Then there exist integers $p_1, p_2 \in [0, \tau^{-1}T]$ such that $p_1 \leq p_2$, $p_1 \leq 2L_0$, $p_2 > \tau^{-1}T - 2L_1$ and that for all integers $i = p_1, \dots, p_2 - 1$,

$$\max\{|x(i\tau + t) - x_f(t)| :\ t \in [0, \tau]\} \leq \varepsilon.$$

Moreover if $|x(0) - x_f(0)| \leq \delta_0$, then $p_1 = 0$ and if $|x(\lfloor \tau^{-1}T \rfloor \tau) - x_f(0)| \leq \delta_0$, then $p_2 = [\tau^{-1}T]$.

Theorem 11 *Let $\varepsilon \in (0, 1)$, $M_0, M_1 > 0$. Then there exist an integer $L \geq 1$, $\delta \in (0, \varepsilon)$ and a neighborhood $\mathcal{U}$ of f in $\mathcal{M}$ such that for each $T > 2L\tau$, each $g \in \mathcal{U}$ and each $(x, u) \in X(A, B, 0, T)$ which satisfies for each $S \in [0, T - L\tau]$,*

$$I^g(S, S + L\tau, x, u) \leq \sigma(g, x(S), x(S + L\tau), S, S + L\tau) + \delta$$

and satisfies at least one of the following conditions below

(a) $|x(0)|,\ |x(T)| \leq M_0$, $I^g(0, T, x, u) \leq \sigma(g, x(0), x(T), 0, T) + M_1$;
(b) $|x(0)| \leq M_0$, $I^g(0, T, x, u) \leq \sigma(g, x(0), 0, T) + M_1$;
(c) $I^g(0, T, x, u) \leq \sigma(g, 0, T) + M_1$

there exist integers $p_1 \in [0, L]$, $p_2 \in [\lfloor \tau^{-1}T \rfloor - L, \tau^{-1}T]$ such that for all integers $i = p_1, \dots, p_2 - 1$,

$|x(i\tau + t) - x_f(t)| \leq \varepsilon$ for all $t \in [0, \tau]$.

Moreover if $|x(0) - x_f(0)| \le \delta$, *then* $p_1 = 0$ *and if* $|x(\lfloor \tau^{-1}T \rfloor \tau) - x_f(0)| \le \delta$, *then* $p_2 = \lfloor \tau^{-1}T \rfloor$.

Denote by Card(A) the cardinality of the set A.

Theorem 12 *Let* $\varepsilon \in (0, 1)$, $M_0, M_1 > 0$. *Then there are an integer* $L \ge 1$ *and a neighborhood* $\mathcal{U}$ *of* f *in* $\mathcal{M}$ *such that for each* $T > L\tau$, *each* $g \in \mathcal{U}$ *and each*

$$(x, u) \in X(A, B, 0, T)$$

which satisfies at least one of the following conditions below

(a) $|x(0)|,\ |x(T)| \le M_0,\ I^g(0, T, x, u) \le \sigma(g, x(0), x(T), 0, T) + M_1$;

(b) $|x(0)| \le M_0,\ I^g(0, T, x, u) \le \sigma(g, x(0), 0, T) + M_1$;

(c) $I^g(0, T, x, u) \le \sigma(g, 0, T) + M_1$

the following inequality holds:

Card$(\{i \in \{0, \dots, \lfloor \tau^{-1}T \rfloor - 1\} :$
$\max\{|x(i\tau + t) - x_f(t)| : t \in [0, \tau]\} > \varepsilon\}) \le L$.

3 Structure of Solutions in the Regions Close to the End Points

In this section we state results obtained in [26] which describe the structure of solutions of problems (P_1), (P_2) and (P_3) in the regions close to the end points. Combined with the turnpike results of Sect. 2 they provide the full description of the structure of their solutions. We use the notation, definitions and assumptions introduced in Sects. 1 and 2.

By Theorem 2 for each $z \in R^n$ there exists a unique (f, A, B)-overtaking optimal pair $(\xi^{(z)}, \eta^{(z)}) \in X(A, B, 0, \infty)$ such that $\xi^{(z)}(0) = z$. Let $z \in R^n$. Set

$$\pi^f(z) = \liminf_{T \to \infty,\, T \in \mathbf{Z}} [I^f(0, T\tau, \xi^{(z)}, \eta^{(z)}) - T\tau\mu(f)]. \tag{3.1}$$

In view of Theorems 1, 2 and 8, $\pi^f(z)$ is a finite number. Definition (3.1) and the definition of (f, A, B)-overtaking optimal pairs imply the following result.

Proposition 2 *1. Let* $(x, u) \in X(A, B, 0, \infty)$ *be* (f, A, B)*-good. Then*

$$\pi^f(x(0)) \le \liminf_{T \to \infty,\, T \in \mathbf{Z}} [I^f(0, T\tau, x, u) - T\tau\mu(f)]$$

and for each pair of integers $S > T \ge 0$,

$$\pi^f(x(T\tau)) \le I^f(T\tau, S\tau, x, u) - (S - T)\tau\mu(f) + \pi^f(x(S\tau)). \tag{3.2}$$

2. Let $S > T \ge 0$ *be integers and* $(x, u) \in X(A, B, T\tau, S\tau)$. *Then (3.2) holds.*

The next result follows from definition (3.1).

Proposition 3 *Let* $(x, u) \in X(A, B, 0, \infty)$ *be* (f, A, B)*-overtaking optimal. Then for each pair of integers* $S > T \geq 0$,

$$\pi^f(x(T\tau)) = I^f(T\tau, S\tau, x, u) - (S - T)\tau\mu(f) + \pi^f(x(S\tau)).$$

Theorems 2–4 and (3.1), (1.3) imply the following result.

Proposition 4 $\pi^f(x_f(0)) = 0$.

The following result is proved in Chap. 2 of [26].

Proposition 5 *The function* π^f *is continuous at* $x_f(0)$.

Proposition 6 *Let* $(x, u) \in X(A, B, 0, \infty)$ *be* (f, A, B)*-overtaking optimal. Then*

$$\pi^f(x(0)) = \lim_{T\to\infty,\, T\in \mathbf{Z}} [I^f(0, T\tau, x, u) - T\tau\mu(f)].$$

Proposition 7 *The function* π^f *is strictly convex and continuous.*

Proposition 8 *For each* $M > 0$ *the set* $\{x \in R^n : \pi^f(x) \leq M\}$ *is bounded.*

Set

$$\inf(\pi^f) = \inf\{\pi^f(z) : z \in R^n\}. \tag{3.3}$$

By Propositions 7 and 8, $\inf(\pi^f)$ is finite and there exists a unique $\theta_f \in R^n$ such that $\pi^f(\theta_f) = \inf(\pi^f)$.

Proposition 9 *Let* $(x, u) \in X(A, B, 0, \infty)$ *be* (f, A, B)*-good such that for all integers* $T > 0$,
$I^f(0, T\tau, x, u) - T\tau\mu(f) = \pi^f(x(0)) - \pi^f(x(T\tau))$.
Then $(x, u) \in X(A, B, 0, \infty)$ *is* (f, A, B)*-overtaking optimal.*

Consider a linear control system

$$x'(t) = -Ax(t) - Bu(t),\ x(0) = x_0$$

which is also controllable. There exists a Borel measurable function $\bar{f} : [0, \infty) \times R^n \times R^m \to R^1$ such that for all $(x, u) \in R^n \times R^m$,

$$\bar{f}(t + \tau, x, u) = \bar{f}(t, x, u) \text{ for all } t \geq 0,$$

$$\bar{f}(t, x, u) = f(\tau - t, x, u) \text{ for all } t \in [0, \tau]. \tag{3.4}$$

Evidently, $\bar{f}$ satisfies assumption (A). For $\bar{f}$ we use all the notation and definitions introduced for f. It is clear that all the results obtained for the triplet (f, A, B) also hold for the triplet $(\bar{f}, -A, -B)$.

Assume that integers $S_2 > S_1 \geq 0$ and that $(x, u) \in X(A, B, S_1\tau, S_2\tau)$. For all $t \in [S_1\tau, S_2\tau]$ set

$$\bar{x}(t) = x(S_2\tau - t + S_1\tau),\ \bar{u}(t) = u(S_2\tau - t + S_1\tau). \tag{3.5}$$

In view of (3.5) for a. e. $t \in [S_1\tau, S_2\tau]$,

$$\bar{x}'(t) = -x'(S_2\tau - t + S_1\tau) = -Ax(S_2\tau - t + S_1\tau) - Bu(S_2\tau - t + S_1\tau)$$

$$= -A\bar{x}(t) - B\bar{u}(t)$$

and $(\bar{x}, \bar{u}) \in X(-A, -B, S_1\tau, S_2\tau)$. By (3.4) and (3.5),

$$\int_{S_1\tau}^{S_2\tau} \bar{f}(t, \bar{x}(t), \bar{u}(t))dt = \int_{S_1\tau}^{S_2\tau} \bar{f}(t, x(S_2\tau - t + S_1\tau), u(S_2\tau - t + S_1\tau))dt$$

$$= \int_{S_1\tau}^{S_2\tau} f(S_2\tau - t + S_1\tau, x(S_2\tau - t + S_1\tau), u(S_2\tau - t + S_1\tau))dt$$

$$= \int_{S_1\tau}^{S_2\tau} f(t, x(t), u(t))dt. \tag{3.6}$$

For each pair $T_2 > T_1 \geq 0$ and each $(x, u) \in X(-A, -B, T_1, T_2)$ set

$$I^{\bar{f}}(T_1, T_2, x, u) = \int_{T_1}^{T_2} \bar{f}(t, x(t), u(t))dt.$$

For each $y, z \in R^n$ and each $T > 0$ set

$$\sigma_-(\bar{f}, y, z, T) = \inf\{I^{\bar{f}}(0, T, x, u) :$$

$$(x, u) \in X(-A, -B, 0, T) \text{ and } x(0) = y,\ x(T) = z\},$$

$$\sigma_-(\bar{f}, y, T) = \inf\{I^{\bar{f}}(0, T, x, u) :\ (x, u) \in X(-A, -B, 0, T) \text{ and } x(0) = y\},$$

$$\widehat{\sigma}_-(\bar{f}, z, T) = \inf\{I^{\bar{f}}(0, T, x, u) :\ (x, u) \in X(-A, -B, 0, T) \text{ and } x(T) = z\},$$

$$\sigma_-(\bar{f}, T) = \inf\{I^{\bar{f}}(0, T, x, u) :\ (x, u) \in X(-A, -B, 0, T)\}. \tag{3.7}$$

Relations (3.5) and (3.6) imply the following result.

Proposition 10 *Let $S_2 > S_1 \geq 0$ be integers, $M \geq 0$ and that*

$$(x_i, u_i) \in X(A, B, S_1\tau, S_2\tau),\ i = 1, 2.$$

Then $I^f(S_1\tau, S_2\tau, x_1, u_1) \ge I^f(S_1\tau, S_2\tau, x_2, u_2) - M$ *if and only if*

$$I^{\bar{f}}(S_1\tau, S_2\tau, \bar{x}_1, \bar{u}_1) \ge I^{\bar{f}}(S_1\tau, S_2\tau, \bar{x}_2, \bar{u}_2) - M.$$

Proposition 10 implies the following result.

Proposition 11 *Let* $S_2 > S_1 \ge 0$ *be integers and*

$$(x, u) \in X(A, B, S_1\tau, S_2\tau).$$

Then the following assertion holds:

$$I^f(S_1\tau, S_2\tau, x, u) \le \sigma(f, (S_2 - S_1)\tau) + M$$

$$\text{if and only if } I^{\bar{f}}(S_1\tau, S_2\tau, \bar{x}, \bar{u}) \le \sigma_-(\bar{f}, (S_2 - S_1)\tau) + M;$$

$$I^f(S_1\tau, S_2\tau, x, u) \le \sigma(f, x(S_1\tau), x(S_2\tau), (S_2 - S_1)\tau) + M$$

$$\text{if and only if } I^{\bar{f}}(S_1\tau, S_2\tau, \bar{x}, \bar{u}) \le \sigma_-(\bar{f}, \bar{x}(S_1\tau), \bar{x}(S_2\tau), (S_2 - S_1)\tau) + M;$$

$$I^f(S_1\tau, S_2\tau, x, u) \le \sigma(f, x(S_1\tau), (S_2 - S_1)\tau) + M$$

$$\text{if and only if } I^{\bar{f}}(S_1\tau, S_2\tau, \bar{x}, \bar{u}) \le \widehat{\sigma}_-(\bar{f}, \bar{x}(S_2\tau), (S_2 - S_1)\tau) + M;$$

$$I^f(S_1\tau, S_2\tau, x, u) \le \widehat{\sigma}(f, x(S_2\tau), (S_2 - S_1)\tau) + M$$

$$\text{if and only if } I^{\bar{f}}(S_1\tau, S_2\tau, \bar{x}, \bar{u}) \le \sigma_-(\bar{f}, \bar{x}(S_1\tau), (S_2 - S_1)\tau) + M.$$

By Proposition 1, $(x_f, u_f) \in X(A, B, 0, \tau)$ is the unique solution of the minimization problem

$$I^f(0, \tau, x, u) \to \min, \ (x, u) \in X(A, B, 0, \tau) \text{ such that } x(0) = x(\tau).$$

Analogously there exists $(x_{\bar{f}}, u_{\bar{f}}) \in X(-A, -B, 0, \tau)$ which is the unique solution of the minimization problem

$$I^{\bar{f}}(0, \tau, x, u) \to \min, \ (x, u) \in X(-A, -B, 0, \tau) \text{ such that } x(0) = x(\tau).$$

In view of Proposition 10 and (3.6), for all $t \in [0, \tau]$,

$$x_{\bar{f}}(t) = x_f(\tau - t), \ u_{\bar{f}}(t) = u_f(\tau - t), \ \mu(\bar{f}) = \mu(f). \tag{3.8}$$

For each $z \in R^n$, set

$$\pi^{\bar{f}}(z) = \liminf_{T\to\infty,\, T\in\mathbf{Z}} [I^{\bar{f}}(0, T\tau, x, u) - T\tau\mu(f)], \tag{3.9}$$

where $(x, u) \in X(-A, -B, 0, \infty)$ is the unique $(\bar{f}, -A, -B)$-overtaking optimal pair such that $x(0) = z$. Let $(x_*, u_*) \in X(A, B, 0, \infty)$ be the unique (f, A, B)-overtaking optimal pair such that $\pi^f(x_*(0)) = \inf(\pi^f)$ and

$$(\bar{x}_*, \bar{u}_*) \in X(-A, -B, 0, \infty)$$

be the unique $(\bar{f}, -A, -B)$-overtaking optimal pair such that

$$\pi^{\bar{f}}(\bar{x}_*(0)) = \inf(\pi^{\bar{f}}).$$

The following three theorems obtained in Chap. 2 of [26] describe the structure of solutions of problems (P_1), (P_2) and (P_3) in the regions closed to the end points.

Theorem 13 *Let $L_0 > 0$ be an integer, $\varepsilon \in (0, 1)$, $M > 0$. Then there exist $\delta > 0$, a neighborhood $\mathcal{U}$ of f in $\mathcal{M}$ and an integer $L_1 > L_0$ such that for each integer $T \geq L_1$, each $g \in \mathcal{U}$ and each $(x, u) \in X(A, B, 0, T\tau)$ which satisfies*

$$|x(0)| \leq M,\ I^g(0, T\tau, x, u) \leq \sigma(g, x(0), 0, T\tau) + \delta$$

the following inequality holds:

$$|x(T\tau - t) - \bar{x}_*(t)| \leq \varepsilon \text{ for all } t \in [0, L_0\tau].$$

Theorem 14 *Let $L_0 > 0$ be an integer, $\varepsilon > 0$. Then there exist $\delta > 0$, a neighborhood $\mathcal{U}$ of f in $\mathcal{M}$ and an integer $L_1 > L_0$ such that for each integer $T \geq L_1$, each $g \in \mathcal{U}$ and each $(x, u) \in X(A, B, 0, T\tau)$ which satisfies*

$$I^g(0, T\tau, x, u) \leq \sigma(g, 0, T\tau) + \delta$$

the following inequalities hold for all $t \in [0, L_0\tau]$:

$$|x(T\tau - t) - \bar{x}_*(t)| \leq \varepsilon,\ |x(t) - x_*(t)| \leq \varepsilon.$$

Theorem 15 *Let $L_0 > 0$ be an integer, $\varepsilon > 0$, $M_0 > 0$. Then there exist $\delta > 0$, a neighborhood $\mathcal{U}$ of f in $\mathcal{M}$ and an integer $L_1 > L_0$ such that for each integer $T \geq L_1$, each $g \in \mathcal{U}$ and each $(x, u) \in X(A, B, 0, T\tau)$ which satisfies*

$$|x(0)|,\ |x(T\tau)| \leq M_0,\ I^g(0, T\tau, x, u) \leq \sigma(g, x(0), x(T\tau), 0, T\tau) + \delta$$

the inequalities

$$|x(T\tau - t) - \bar{\xi}(t)| \leq \varepsilon, \ |x(t) - \xi(t)| \leq \varepsilon$$

hold for all $t \in [0, L_0\tau]$, *where* $(\xi, \eta) \in X(A, B, 0, \infty)$ *is the unique* (f, A, B)-*overtaking optimal pair such that* $\xi(0) = x(0)$ *and*

$$(\bar{\xi}, \bar{\eta}) \in X(-A, -B, 0, \infty)$$

is the unique $(\bar{f}, -A, -B)$-*overtaking optimal pair such that* $\bar{\xi}(0) = x(T\tau)$.

4 Bolza Optimal Control Problems

We use the notation, definitions and assumptions introduced in Sects. 1–3. Recall that $f : [0, \infty) \times R^n \times R^m \to R^1$ is a Borel measurable function which satisfy assumptions (A). Let $a > 0$ and $\psi : [0, \infty) \to [0, \infty)$ be an increasing function such that

$$\lim_{t \to \infty} \psi(t) = \infty. \tag{4.1}$$

We suppose that for all $(t, x, u) \in [0, \infty) \times R^n \times R^m$,

$$f(t, x, u) \geq \max\{\psi(|x|), \ \psi(|u|)|u|\} - a. \tag{4.2}$$

We consider the complete metric space of Borel measurable functions $\mathcal{M}$ introduced in Sect. 2.

Let $a_1 > 0$ and $k \geq 1$ be an integer. Denote by $\mathfrak{A}_k$ the set of all lower semicontinuous functions $h : R^k \to R^1$ which are bounded on bounded subsets of R^k and satisfy

$$h(z) \geq -a_1 \text{ for all } z \in R^k. \tag{4.3}$$

We equip the set $\mathfrak{A}_k$ with the uniformity which is determined by the following base:

$$E_k(N, \varepsilon) = \{(h_1, h_2) \in \mathfrak{A}_k \times \mathfrak{A}_k : \ |h_1(z) - h_2(z)| \leq \varepsilon$$

$$\text{for each } z \in R^k \text{ satisfying } |z| \leq N\}, \tag{4.4}$$

where $N > 0, \varepsilon > 0$. It is not difficult to see that the uniform space $\mathfrak{A}_k$ is metrizable and complete.

Let $g \in \mathcal{M}, h, \xi \in \mathfrak{A}_n, H \in \mathfrak{A}_{2n}, y \in R^n$ and $\infty > T_2 > T_1 \geq 0$. We consider the following optimal control problems

$$I^g(T_1, T_2, x, u) + h(x(T_2)) \to \min,$$

$$(x, u) \in X(A, B, T_1, T_2) \text{ such that } x(T_1) = y$$

and

$$I^g(T_1, T_2, x, u) + H(x(T_1), x(T_2)) \to \min,$$

$$(x, u) \in X(A, B, T_1, T_2)$$

and define

$$\sigma(g, h, y, T_1, T_2) = \inf\{I^g(T_1, T_2, x, u) + h(x(T_2)) :$$

$$(x, u) \in X(A, B, T_1, T_2) \text{ and } x(T_1) = y\}, \tag{4.5}$$

$$\widehat{\sigma}(g, \xi, z, T_1, T_2) = \inf\{I^g(T_1, T_2, x, u) + \xi(x(T_1)) :$$

$$(x, u) \in X(A, B, T_1, T_2) \text{ and } x(T_2) = z\}, \tag{4.6}$$

$$\sigma(g, h, \xi, T_1, T_2) = \inf\{I^g(T_1, T_2, x, u) + h(x(T_2)) + \xi(x(T_1)) :$$

$$(x, u) \in X(A, B, T_1, T_2)\}, \tag{4.7}$$

$$\sigma(g, H, T_1, T_2) = \inf\{I^g(T_1, T_2, x, u) + H(x(T_1), x(T_2)) :$$

$$(x, u) \in X(A, B, T_1, T_2)\}. \tag{4.8}$$

Since every $g \in \mathcal{M}$ and every $h \in \mathfrak{A}_k$, $k = n, 2n$ are bounded on bounded sets it follows from Proposition 2.28 of [26], (2.3) and (4.3) that all the values defined above are finite. Set

$$M_* = \max\{\sup\{|x_*(t)| : t \in [0, \infty)\}, \ \sup\{|\bar{x}_*(t)| : t \in [0, \infty)\}. \tag{4.9}$$

We prove the following turnpike results for our Bolza optimal control problems which show that the turnpike phenomenon, for approximate solutions on large intervals, is stable under small perturbations of the objective functions. In Theorems 16 and 17 we consider problems on intervals $[0, T\tau]$ where T is a natural number while in Theorems 18 and 19 the Bolza problems are considered on intervals $[0, T]$, where T is a sufficiently large positive number.

Theorem 16 *Let* $\varepsilon \in (0, 1)$, $M_0, M_1, M_2 > 0$. *Then there exist an integer* $L \geq 1$ *and a neighborhood* $\mathcal{U}$ *of* f *in* $\mathcal{M}$ *such that for each integer* $T > L$, *each* $g \in \mathcal{U}$, *each* $h \in \mathfrak{A}_n$ *and each* $\xi \in \mathfrak{A}_{2n}$ *which satisfy*

$$h(z) \leq M_2 \textit{ for all } z \in R^n \textit{ satisfying } |z| \leq M_* + 1, \tag{4.10}$$

$$\xi(z) \le M_2 \textit{ for all } z = (z_1, z_2) \in R^n \times R^n \textit{ satisfying } |z_i| \le M_* + 1, \ \ i = 1, 2 \tag{4.11}$$

and each $(x, u) \in X(A, B, 0, T\tau)$ *which satisfies at least one of the following conditions below*

$$(a)\ |x(0)| \le M_0,$$

$$I^g(0, T\tau, x, u) + h(x(T\tau)) \le \sigma(g, h, x(0), 0, T\tau) + M_1;$$

$$(b)\ I^g(0, T\tau, x, u) + \xi(x(0), x(T\tau)) \le \sigma(g, \xi, 0, T\tau) + M_1$$

the following inequality holds:

$$Card(\{i \in \{0, \dots, T-1\} :\ \max\{|x(i\tau + t) - x_f(t)| :\ t \in [0, \tau]\} > \varepsilon\}) \le L.$$

Theorem 17 *Let* $\varepsilon \in (0, 1),\ M_0, M_1, M_2 > 0$. *Then there exist an integer* $L \ge 1$, $\delta \in (0, \varepsilon)$ *and a neighborhood* $\mathscr{U}$ *of* f *in* $\mathscr{M}$ *such that for each integer* $T > 2L$, *each* $g \in \mathscr{U}$, *each* $h \in \mathfrak{A}_n$ *and each* $\xi \in \mathfrak{A}_{2n}$ *which satisfy*

$$h(z) \le M_2 \textit{ for all } z \in R^n \textit{ satisfying } |z| \le M_* + 1,$$

$$\xi(z) \le M_2 \textit{ for all } z = (z_1, z_2) \in R^n \times R^n \textit{ satisfying } |z_i| \le M_* + 1, \ \ i = 1, 2$$

and each $(x, u) \in X(A, B, 0, T\tau)$ *which satisfies for each* $S \in [0, T\tau - L\tau]$,

$$I^g(S, S + L\tau, x, u) \le \sigma(g, x(S), x(S + L\tau), S, S + L\tau) + \delta$$

and satisfies at least one of the following conditions below

$$(a)\ |x(0)| \le M_0,$$

$$I^g(0, T\tau, x, u) + h(x(T\tau)) \le \sigma(g, h, x(0), 0, T\tau) + M_1;$$

$$(b)\ I^g(0, T\tau, x, u) + \xi(x(0), x(T\tau)) \le \sigma(g, \xi, 0, T\tau) + M_1$$

there exist integers $p_1 \in [0, L],\ p_2 \in [T - L, T]$ *such that for all integers* $i = p_1, \dots, p_2 - 1$,

$$|x(i\tau + t) - x_f(t)| \le \varepsilon \textit{ for all } t \in [0, \tau].$$

Moreover if $|x(0) - x_f(0)| \le \delta$, *then* $p_1 = 0$ *and if* $|x(T\tau) - x_f(0)| \le \delta$, *then* $p_2 = T$.

Theorem 18 *Let* $\varepsilon \in (0, 1),\ M_0, M_1 > 0$. *Then there exists* $\tilde{M} > 0$ *such that for each* $M_2 > 0$ *there exist an integer* $L \ge 1$ *and a neighborhood* $\mathscr{U}$ *of* f *in* $\mathscr{M}$ *such that for each* $T > L\tau$, *each* $g \in \mathscr{U}$, *each* $h \in \mathfrak{A}_n$ *which satisfies*

$$h(z) \le M_2 \text{ for all } z \in R^n \text{ satisfying } |z| \le \tilde{M}, \tag{4.12}$$

each $\xi \in \mathfrak{A}_{2n}$ *which satisfy*

$$\xi(z) \le M_2 \text{ for all } z = (z_1, z_2) \in R^n \times R^n \text{ satisfying } |z_i| \le \tilde{M},\ i = 1, 2 \tag{4.13}$$

and each $(x, u) \in X(A, B, 0, T)$ *which satisfies at least one of the following conditions below*

$$(a)\ |x(0)| \le M_0,$$

$$I^g(0, T, x, u) + h(x(T)) \le \sigma(g, h, x(0), 0, T) + M_1;$$

$$(b)\ I^g(0, T, x, u) + \xi(x(0), x(T)) \le \sigma(g, \xi, 0, T) + M_1$$

the following inequality holds:

$$Card(\{i \in \{0, \dots, \lfloor \tau^{-1}T \rfloor - 1\} :$$

$$\max\{|x(i\tau + t) - x_f(t)| : t \in [0, \tau]\} > \varepsilon\}) \le L.$$

Theorem 19 *Let* $\varepsilon \in (0, 1)$, $M_0, M_1 > 0$ *and* $\tilde{M}$ *be as guaranteed by Theorem 18. Let* $M_2 > 0$. *Then there exist an integer* $L \ge 1$, $\delta \in (0, \varepsilon)$ *and a neighborhood* $\mathscr{U}$ *of* f *in* $\mathscr{M}$ *such that for each* $T > L\tau$, *each* $g \in \mathscr{U}$, *each* $h \in \mathfrak{A}_n$ *which satisfy*

$$h(z) \le M_2 \text{ for all } z \in R^n \text{ satisfying } |z| \le \tilde{M},$$

each $\xi \in \mathfrak{A}_{2n}$ *which satisfy*

$$\xi(z) \le M_2 \text{ for all } z = (z_1, z_2) \in R^n \times R^n \text{ satisfying } |z_i| \le \tilde{M},\ i = 1, 2$$

and each $(x, u) \in X(A, B, 0, T)$ *which satisfies for each* $S \in [0, T - L\tau]$,

$$I^g(S, S + L\tau, x, u) \le \sigma(g, x(S), x(S + L\tau), S, S + L\tau) + \delta$$

and satisfies at least one of the following conditions below

$$(a)\ |x(0)| \le M_0,$$

$$I^g(0, T, x, u) + h(x(T)) \le \sigma(g, h, x(0), 0, T) + M_1;$$

$$(b)\ I^g(0, T, x, u) + \xi(x(0), x(T)) \le \sigma(g, \xi, 0, T) + M_1$$

there exist integers $p_1 \in [0, L]$, $p_2 \in [\lfloor \tau^{-1}T \rfloor - L, \tau^{-1}T]$ *such that for all integers* $i = p_1, \dots, p_2 - 1$,

$$|x(i\tau + t) - x_f(t)| \le \varepsilon \text{ for all } t \in [0, \tau].$$

Moreover if $|x(0) - x_f(0)| \le \delta$, *then* $p_1 = 0$ *and if* $|x(\lfloor \tau^{-1} T \rfloor \tau) - x_f(0)| \le \delta$, *then* $p_2 = \lfloor \tau^{-1} T \rfloor$.

Let $h \in \mathfrak{A}_n$ and $\xi \in \mathfrak{A}_{2n}$. Define

$$\psi_\xi(z_1, z_2) = \pi^f(z_1) + \pi^{\bar{f}}(z_2) + \xi(z_1, z_2), \; z_1, z_2 \in R^n. \tag{4.14}$$

Propositions 7 and 8 imply the following results.

Proposition 12 *The function* $\pi^f + h$ *is lower semicontinuous, for every* $M > 0$ *the set* $\{x \in R^n : (\pi^f + h)(x) \le M\}$ *is bounded,* $\inf(\pi^f + h)$ *is finite and the function* $\pi^f + h$ *has a point of minimum.*

Proposition 13 *The function* ψ_ξ *is lower semicontinuous, for every* $M > 0$ *the set* $\{(z_1, z_2) \in R^n \times R^n : \psi_\xi(z_1, z_2) \le M\}$ *is bounded and the function* ψ_ξ *has a point of minimum.*

In this paper we also prove the following two stability results for our Bolza optimal control problems. They show that the convergence of approximate solutions on large intervals, in the regions close to the end points, is stable under small perturbations of the objective functions.

Theorem 20 *Let* $L_0 \ge 1$ *be an integer,* $h \in \mathfrak{A}_n$, $\varepsilon \in (0, 1)$, $M > 0$. *Then there exist* $\delta > 0$, *a neighborhood* $\mathcal{U}$ *of* f *in* $\mathcal{M}$, *a neighborhood* $\mathcal{V}$ *of* h *in* $\mathfrak{A}_n$ *and an integer* $L_1 > L_0$ *such that for each integer* $T \ge L_1$, *each* $g \in \mathcal{U}$, *each* $\xi \in \mathcal{V}$ *and each* $(x, u) \in X(A, B, 0, T\tau)$ *which satisfies*

$$|x(0)| \le M,$$

$$I^g(0, T\tau, x, u) + \xi(x(T\tau)) \le \sigma(g, \xi, x(0), 0, T\tau) + \delta$$

there exists an $(\bar{f}, -A, -B)$*-overtaking optimal pair*

$$(\bar{x}_h, \bar{u}_h) \in X(-A, -B, 0, \infty)$$

such that

$$(\pi^{\bar{f}} + h)(\bar{x}_h(0)) = \inf(\pi^{\bar{f}} + h),$$

$$|x(T\tau - t) - \bar{x}_h(t)| \le \varepsilon \text{ for all } t \in [0, L_0\tau].$$

Theorem 21 *Let* $L_0 \ge 1$ *be an integer,* $h \in \mathfrak{A}_{2n}$, $\varepsilon \in (0, 1)$. *Then there exist* $\delta > 0$, *a neighborhood* $\mathcal{U}$ *of* f *in* $\mathcal{M}$, *a neighborhood* $\mathcal{V}$ *of* h *in* $\mathfrak{A}_{2n}$ *and an integer* $L_1 > L_0$ *such that for each integer* $T \ge L_1$, *each* $g \in \mathcal{U}$, *each* $\xi \in \mathcal{V}$ *and each* $(x, u) \in X(A, B, 0, T\tau)$ *which satisfies*

$$I^g(0, T\tau, x, u) + \xi(x(0), x(T\tau)) \le \sigma(g, \xi, 0, T\tau) + \delta$$

there exist an (f, A, B)*-overtaking optimal pair* $(x_*, u_*) \in X(A, B, 0, \infty)$ *and an* $(\bar{f}, -A, -B)$*-overtaking optimal pair* $(\bar{x}_*, \bar{u}_*) \in X(-A, -B, 0, \infty)$ *such that*

$$\psi_h(x_*(0), \bar{x}_*(0)) = \inf(\psi_h)$$

and for all $t \in [0, L_0\tau]$,

$$|x(t) - x_*(t)| \le \varepsilon, \ |x(T\tau - t) - \bar{x}_*(t)| \le \varepsilon.$$

5 Auxiliary Results

In the sequel we use the following auxiliary results.

Proposition 14 (Proposition 2.35 of [26]) *Let* $M_0 > 0$. *Then there exists* $M > 0$ *such that for each* $T \ge 3\tau$ *and each* $y, z \in R^n$ *satisfying* $|y|, |z| \le M_0$,

$$\sigma(f, y, z, T) \le T\mu(f) + M.$$

Proposition 15 (Proposition 2.40 of [26]) *Let* $M_1 > 0$, $0 < \tau_0 < \tau_1$. *Then there exists* $M_2 > 0$ *such that for each* $g \in \mathcal{M}$, *each* $T_2 > T_1 \ge 0$ *satisfying*

$$T_2 - T_1 \in [\tau_0, \tau_1]$$

and each $(x, u) \in X(A, B, T_1, T_2)$ *satisfying* $I^g(T_1, T_2, x, u) \le M_1$ *the following inequality holds:* $|x(t)| \le M_2$ *for all* $t \in [T_1, T_2]$.

Proposition 16 (Proposition 2.41 of [26]) *Let* $0 < c_1 < c_2, \ D, \varepsilon > 0$. *Then there exists a neighborhood* V *of* f *in* $\mathcal{M}$ *such that for each* $g \in V$, *each* $T_2 > T_1 \ge 0$ *satisfying* $T_2 - T_1 \in [c_1, c_2]$ *and each* $(x, u) \in X(A, B, T_1, T_2)$ *satisfying*

$$\min\{I^f(T_1, T_2, x, u), I^g(T_1, T_2, x, u)\} \le D$$

the inequality $|I^f(T_1, T_2, x, u) - I^g(T_1, T_2, x, u)| \le \varepsilon$ *holds.*

Proposition 17 (Proposition 6.2.4 of [22], Proposition 2.30 of [26]) *Let* M_1 *and* T *be positive numbers and let* $\mathcal{F}$ *be the set of all* $(x, u) \in X(A, B, 0, T)$ *satisfying* $I^f(0, T, x, u) \le M_1$. *Then for every sequence* $\{(x_i, u_i)\}_{i=1}^{\infty} \subset \mathcal{F}$ *there exist a subsequence* $\{(x_{i_k}, u_{i_k})\}_{k=1}^{\infty}$ *and* $(x, u) \in \mathcal{F}$ *such that* $x_{i_k}(t) \to x(t)$ *as* $k \to \infty$ *uniformly in* $[0, T]$, $x'_{i_k} \to x'$ *as* $k \to \infty$ *weakly in* $L^1(R^n; (0, T))$, *and* $u_{i_k} \to u$ *as* $k \to \infty$ *weakly in* $L^1(R^m; (0, T))$.

For each $y, z \in R^n$ define

$$v(y, z) = \inf\{I^f(0, \tau, x, u) : (x, u) \in X(A, B, 0, \tau)$$

$$\text{such that } x(0) = y,\ x(\tau) = z\}. \tag{5.1}$$

It was shown in Sect. 6.2 of [22] that the function v is convex, satisfies

$$-\infty < v(y, z) < \infty \text{ for each } y, z \in R^n,$$

$$v(y, z) \to \infty \text{ as } |y| + |z| \to \infty \tag{5.2}$$

and that there exists $z_f \in R^n$ such that

$$v(z_f, z_f) < v(z, z) \text{ for all } z \in R^n \setminus \{z_f\}, \tag{5.3}$$

$$x_f(0) = z_f,\ \mu(f) = \tau^{-1} v(z_f, z_f). \tag{5.4}$$

Proposition 18 (Corollary 6.2.1 of [22]) *Let $x_1, x_2 \in R^n$. Then there is a unique $(x, u) \in X(A, B, 0, \tau)$ such that $x(0) = x_1$, $x(\tau) = x_2$ and $I^f(0, \tau, x, u) = v(x_1, x_2)$.*

6 Proof of Theorem 16

By Theorems 13 and 14, there exist $\delta \in (0, 1)$, a neighborhood $\mathcal{U}_1$ of f in $\mathcal{M}$ and an integer $L_1 > 1$ such that the following properties hold:

(P1) for each integer $T \geq L_1$, each $g \in \mathcal{U}_1$ and each $(x, u) \in X(A, B, 0, T\tau)$ which satisfies

$$|x(0)| \leq M_0,$$

$$I^g(0, T\tau, x, u) \leq \sigma(g, x(0), 0, T\tau) + \delta$$

the inequality $|x(T\tau - t) - \bar{x}_*(t)| \leq 1$ holds for all $t \in [0, \tau]$:

(P2) for each integer $T \geq L_1$, each $g \in \mathcal{U}_1$ and each $(x, u) \in X(A, B, 0, T\tau)$ which satisfies $I^g(0, T\tau, x, u) \leq \sigma(g, 0, T\tau) + \delta$ the inequalities

$$|x(T\tau - t) - \bar{x}_*(t)| \leq 1,\ |x(t) - x_*(t)| \leq 1$$

hold for all $t \in [0, \tau]$.

By Theorem 12, there exist an integer $L > L_1$ and a neighborhood $\mathcal{U} \subset \mathcal{U}_1$ of f in $\mathcal{M}$ such that the following property holds:

(P3) for each integer $T > L$, each $g \in \mathcal{U}$ and each $(x, u) \in X(A, B, 0, T\tau)$ which satisfies at least one of the following conditions below

$$|x(0)| \le M_0,\ I^g(0, T\tau, x, u) \le \sigma(g, x(0), 0, T\tau) + 1 + M_1 + M_2 + a_1;$$

$$I^g(0, T\tau, x, u) \le \sigma(g, 0, T\tau) + 1 + M_1 + 2M_2 + 2a_1$$

we have

$$\mathrm{Card}(\{i \in \{0, \ldots, T-1\} : \max\{|x(i\tau + t) - x_f(t)| : t \in [0, \tau]\} > \varepsilon\}) \le L.$$

Assume that an integer

$$T > L,\ g \in \mathscr{U}, \tag{6.1}$$

$h \in \mathfrak{A}_n, \xi \in \mathfrak{A}_{2n}$ satisfy (4.10) and (4.11), $(x, u) \in X(A, B, 0, T\tau)$ and at least one of the conditions (a) and (b) holds. There exists $(y, v) \in X(A, B, 0, T\tau)$ such that if condition (a) holds, then

$$y(0) = x(0), \tag{6.2}$$

$$I^g(0, T\tau, y, v) \le \sigma(g, x(0), 0, T\tau) + \delta \tag{6.3}$$

and if condition (b) holds, then

$$I^g(0, T\tau, y, v) \le \sigma(g, 0, T\tau) + \delta. \tag{6.4}$$

By (6.1)–(6.4) and properties (P1) and (P2), if condition (a) holds, then

$$|y(T\tau - t) - \bar{x}_*(t)| \le 1,\ t \in [0, \tau], \tag{6.5}$$

and if condition (b) holds, then

$$|y(T\tau - t) - \bar{x}_*(t)| \le 1,\ |y(t) - x_*(t)| \le 1 \tag{6.6}$$

hold for all $t \in [0, \tau]$. It follows from (4.9), (4.10), (6.5) and (6.6) that if condition (a) holds, then

$$|y(T\tau)| \le M_* + 1,\ h(y(T\tau)) \le M_2 \tag{6.7}$$

and if condition (b) holds, then

$$|y(0)|, |y(T\tau)| \le M_* + 1,\ \xi(y(0), y(T\tau)) \le M_2. \tag{6.8}$$

Assume that condition (a) holds. By (4.3), (6.2), (6.3), (6.7) and condition (a),

$$I^g(0, T\tau, x, u) - a_1 \le I^g(0, T\tau, x, u) + h(x(T\tau))$$

$$\le I^g(0, T\tau, y, v) + h(y(T\tau)) + M_1 \le I^g(0, T\tau, y, v) + M_1 + M_2$$

$$\leq \sigma(g, x(0), 0, T\tau) + 1 + M_1 + M_2,$$

$$I^g(0, T\tau, x, u) \leq \sigma(g, x(0), 0, T\tau) + 1 + M_1 + M_2 + a_1.$$

In view of the inequality above, (6.1), condition (a) and property (P3),

$$\mathrm{Card}(\{i \in \{0, \dots, T-1\} : \max\{|x(i\tau + t) - x_f(t)| : t \in [0, \tau]\} > \varepsilon\}) \leq L. \tag{6.9}$$

Assume that condition (b) holds. By (4.3), (6.4), (6.8) and condition (b),

$$I^g(0, T\tau, x, u) - a_1 \leq I^g(0, T\tau, x, u) + \xi(x(0), x(T\tau))$$

$$\leq I^g(0, T\tau, y, v) + \xi(y(0), y(T\tau)) + M_1 \leq I^g(0, T\tau, y, v) + M_1 + M_2$$

$$\leq \sigma(g, 0, T\tau) + 1 + M_1 + M_2,$$

$$I^g(0, T\tau, x, u) \leq \sigma(g, 0, T\tau) + 1 + M_1 + M_2 + a_1.$$

In view of the inequality above, (6.1), condition (b) and property (P3), inequality (6.9) is true. Theorem 16 is proved.

7 Proof of Theorem 17

By Theorem 11, there exist an integer $L_1 \geq 1$, $\delta \in (0, \varepsilon)$ and a neighborhood $\mathscr{U}_1$ of f in $\mathscr{M}$ such that the following property holds:

(P4) for each integer $T > 2L_1$, each $g \in \mathscr{U}_1$ and each

$$(x, u) \in X(A, B, 0, T\tau)$$

which satisfies for each $S \in [0, T\tau - L_1\tau]$,

$$I^g(S, S + L_1\tau, x, u) \leq \sigma(g, x(S), x(S + L_1\tau), S, S + L_1\tau) + \delta$$

and satisfies

$$|x(0) - x_f(0)| \leq \delta, \ |x(T\tau) - x_f(0)| \leq \delta,$$

$$I^g(0, T\tau, x, u) \leq \sigma(g, x(0), x(T\tau), 0, T\tau) + M_1$$

we have for all integers $i = 0, \dots, T-1$,

$$|x(i\tau + t) - x_f(t)| \leq \varepsilon \text{ for all } t \in [0, \tau].$$

By Theorem 16, there exist an integer $L_2 \geq 1$ and a neighborhood $\mathcal{U} \subset \mathcal{U}_1$ of f in $\mathcal{M}$ such that the following property holds:

(P5) for each integer $T > L_2$, each $g \in \mathcal{U}$, each $h \in \mathfrak{A}_n$ and each $\xi \in \mathfrak{A}_{2n}$ which satisfy

$$h(z) \leq M_2 \text{ for all } z \in R^n \text{ satisfying } |z| \leq M_* + 1, \tag{7.1}$$

$$\xi(z) \leq M_2 \text{ for all } z = (z_1, z_2) \in R^n \times R^n \text{ satisfying } |z_i| \leq M_* + 1, \ i = 1, 2 \tag{7.2}$$

and each $(x, u) \in X(A, B, 0, T\tau)$ which satisfies at least one of the following conditions below

(a) $|x(0)| \leq M_0$,

$$I^g(0, T\tau, x, u) + h(x(T\tau)) \leq \sigma(g, h, x(0), 0, T\tau) + M_1;$$

(b) $I^g(0, T\tau, x, u) + \xi(x(0), x(T\tau)) \leq \sigma(g, \xi, 0, T\tau) + M_1$

we have

$$\text{Card}(\{i \in \{0, \dots, T-1\} : \max\{|x(i\tau + t) - x_f(t)| : t \in [0, \tau]\} > \delta\}) < L_2.$$

Choose an integer

$$L \geq 4L_1 + 4L_2. \tag{7.3}$$

Assume that an integer

$$T > 2L, \ g \in \mathcal{U}, \ h \in \mathfrak{A}_n, \ \xi \in \mathfrak{A}_{2n}, \tag{7.4}$$

Equations (7.1) and (7.2) hold and that $(x, u) \in X(A, B, 0, T\tau)$ satisfies for each $S \in [0, T\tau - L\tau]$,

$$I^g(S, S + L\tau, x, u) \leq \sigma(g, x(S), x(S + L\tau), S, S + L\tau) + \delta \tag{7.5}$$

and satisfies at least one of the conditions (a) and (b). Together with property (P5) this implies that there exist integers

$$p_1 \in [0, L_2], \ p_2 \in [T - L_2, T - 1] \tag{7.6}$$

such that

$$|x(p_i\tau) - x_f(0)| \leq \delta, \ i = 1, 2. \tag{7.7}$$

If $|x(0) - x_f(0)| \le \delta$, then we set $p_1 = 0$ and if $|x(T\tau) - x_f(0)| \le \delta$, then we set $p_2 = T$. By (7.3)–(7.5), (7.7) and property (P4), for all integers $i = p_1, \ldots, p_2 - 1$,

$$|x(i\tau + t) - x_f(t)| \le \varepsilon \text{ for all } t \in [0, \tau].$$

Theorem 17 is proved.

8 Proof of Theorem 18

By Theorem 12, there exist an integer $L_0 \ge 4$ and a neighborhood $\mathscr{U}_1$ of f in $\mathscr{M}$ such that the following property holds:

(P6) for each $T > L_0\tau$, each $g \in \mathscr{U}_1$ and each $(x, u) \in X(A, B, 0, T)$ which satisfies at least one of the following conditions below

$$|x(0)| \le M_0, \; I^g(0, T, x, u) \le \sigma(g, x(0), 0, T) + M_1 + 2;$$

$$I^g(0, T, x, u) \le \sigma(g, 0, T) + M_1 + 2$$

we have

$$\text{Card}(\{i \in \{0, \ldots, \lfloor \tau^{-1} T \rfloor - 1\} :$$

$$\max\{|x(i\tau + t) - x_f(t)| : \; t \in [0, \tau]\} > 1\}) \le L_0.$$

By Proposition 14, there exists $M_2 > 0$ such that the following property holds:

(P7) for each $S \ge 3\tau$ and each $z_1, z_2 \in R^n$ satisfying $|z_i| \le M_* + 2, i = 1, 2$,

$$\sigma(f, z_1, z_2, S) \le S\mu(f) + M_2.$$

By Proposition 16, there exists a neighborhood $\mathscr{U}_2$ of f in $\mathscr{M}$ such that the following property holds:

(P8) for each $g \in \mathscr{U}_2$, each $T_2 > T_1 \ge 0$ satisfying $T_2 - T_1 \in [L_0\tau, 4(L_0 + 1)\tau]$ and each $(x, u) \in X(A, B, T_1, T_2)$ satisfying

$$\min\{I^f(T_1, T_2, x, u), I^g(T_1, T_2, x, u)\} \le 4(L_0 + 1)\tau|\mu(f)| + M_2 + 4$$

we have

$$|I^f(T_1, T_2, x, u) - I^g(T_1, T_2, x, u)| \le 1.$$

By Proposition 15, there exists $\tilde{M} > 0$ such that the following property holds:

(P9) for each $g \in \mathscr{M}$, each $T_2 > T_1 \ge 0$ satisfying

$$T_2 - T_1 \in [(L_0 - 1)\tau, 4(L_0 + 1)\tau]$$

and each $(x, u) \in X(A, B, T_1, T_2)$ satisfying

$$I^g(T_1, T_2, x, u) \le 4(L_0 + 1)|\mu(f)|\tau + M_2 + 4$$

we have

$$|x(t)| \le \tilde{M} \text{ for all } t \in [T_1, T_2].$$

Let $M_2 > 0$. By Theorem 12, there exist an integer $L > 8L_0 + 4$ and a neighborhood $\mathscr{U} \subset \mathscr{U}_1 \cap \mathscr{U}_2$ of f in $\mathscr{M}$ such that the following property holds:

(P10) for each $T > L\tau$, each $g \in \mathscr{U}$ and each $(x, u) \in X(A, B, 0, T)$ which satisfies at least one of the following conditions below

$$|x(0)| \le M_0,\ I^g(0, T, x, u) \le \sigma(g, x(0), 0, T) + 1 + M_1 + M_2 + a_1;$$

$$I^g(0, T, x, u) \le \sigma(g, 0, T) + M_1 + 1 + 2M_2 + 2a_1$$

we have

$$\text{Card}(\{i \in \{0, \dots, \lfloor \tau^{-1}T \rfloor - 1\} :$$

$$\max\{|x(i\tau + t) - x_f(t)| : t \in [0, \tau]\} > \varepsilon\}) \le L.$$

Assume that

$$T > L\tau,\ g \in \mathscr{U},\ h \in \mathfrak{A}_n,\ \xi \in \mathfrak{A}_{2n}, \tag{8.1}$$

Equations (4.12) and (4.13) hold, $(x, u) \in X(A, B, 0, T)$ and at least one of the conditions (a) and (b) of Theorem 18 holds. There exists $(y, v) \in X(A, B, 0, T)$ such that if condition (a) holds, then

$$y(0) = x(0),\ I^g(0, T, y, v) \le \sigma(g, x(0), 0, T) + 1 \tag{8.2}$$

and in condition (b) holds, then

$$I^g(0, T, y, v) \le \sigma(g, 0, T) + 1. \tag{8.3}$$

By (8.1)–(8.3), property (P6) and conditions (a) and (b) there exist

$$i_1 \in \{L_0 - 1, \dots, 2L_0 - 1\},\ i_2 \in \{\lfloor \tau^{-1}T \rfloor - 2L_0 - 1, \dots, \lfloor \tau^{-1}T \rfloor - L_0 - 1\} \tag{8.4}$$

such that

$$|y(i_1\tau + t) - x_f(t)| \le 1,\ |y(i_2\tau + t) - x_f(t)| \le 1,\ t \in [0, \tau]. \tag{8.5}$$

It follows from (4.9) and (8.5) that

$$|y(i_1\tau)|,\ |y(i_2\tau)| \le M_* + 1. \tag{8.6}$$

It is not difficult to see that

$$I^g(i_2\tau, T, y, v) \le \sigma(g, y(i_2\tau), i_2\tau, T) + 1 \tag{8.7}$$

and that if the case (b) holds, then

$$I^g(0, i_1\tau, y, v) \le \widehat{\sigma}(g, y(i_1\tau), 0, i_1\tau) + 1. \tag{8.8}$$

Property (P7), (8.4) and (8.6) imply that

$$\sigma(f, y(i_2\tau), 0, i_2\tau, T) \le (T - i_2\tau)\mu(f) + M_2 \tag{8.9}$$

and

$$\sigma(f, 0, y(i_1\tau), 0, i_1\tau) \le i_1\tau\mu(f) + M_2. \tag{8.10}$$

It follows from (8.1), (8.4), (8.9), (8.10) and property (P8) that

$$\sigma(g, y(i_2\tau), 0, i_2\tau, T) \le (T - i_2\tau)\mu(f) + M_2 + 2, \tag{8.11}$$

$$\sigma(g, 0, y(i_1\tau), 0, i_1\tau) \le i_1\tau\mu(f) + M_2 + 2. \tag{8.12}$$

In view of (8.7), (8.8), (8.11) and (8.12),

$$I^g(i_2\tau, T, y, v) \le (T - i_2\tau)\mu(f) + M_2 + 3 \tag{8.13}$$

and if the case (b) holds, then

$$I^g(0, i_1\tau, y, v) \le i_1\tau\mu(f) + M_2 + 3. \tag{8.14}$$

Property (P9), (8.4), (8.13) and (8.14) imply that

$$|y(T)| \le \tilde{M} \tag{8.15}$$

and if condition (b) holds, then

$$|y(0)| \le \tilde{M}. \tag{8.16}$$

Assume that condition (a) holds. By (4.3), (4.12), (8.2), (8.15) and condition (a),

$$I^g(0, T, x, u) - a_1 \le I^g(0, T, x, u) + h(x(T))$$

$$\le I^g(0, T, y, v) + h(y(T)) + M_1 \le I^g(0, T, y, v) + M_1 + M_2$$

$$\leq \sigma(g, x(0), 0, T) + 1 + M_1 + M_2,$$

$$I^g(0, T, x, u) \leq \sigma(g, x(0), 0, T) + M_1 + M_2 + 1 + a_1.$$

In view of the inequality above, (8.1), condition (a) and property (P10),

$$\text{Card}(\{i \in \{0, \dots, \lfloor \tau^{-1} T \rfloor - 1\} :$$

$$\max\{|x(i\tau + t) - x_f(t)| : t \in [0, \tau]\} > \varepsilon\}) \leq L. \tag{8.17}$$

Assume that condition (b) holds. By (4.3), (4.13), (8.3), (8.15) and (8.16),

$$I^g(0, T, x, u) - a_1 \leq I^g(0, T, x, u) + \xi(x(0), x(T))$$

$$\leq I^g(0, T, y, v) + \xi(y(0), y(T\tau)) + M_1 \leq \sigma(g, 0, T) + 1 + M_1 + M_2,$$

$$I^g(0, T, x, u) \leq \sigma(g, 0, T) + 1 + M_1 + M_2 + a_1.$$

In view of the inequality above, (8.1) and property (P10), inequality (8.17) is true. Theorem 18 is proved.

9 Proof of Theorem 19

By Theorem 11, there exist an integer $L_1 \geq 1$, $\delta \in (0, \varepsilon)$ and a neighborhood $\mathscr{U}_1$ of f in $\mathscr{M}$ such that the following property holds:

(P11) for each integer $T > 2L_1$, each $g \in \mathscr{U}_1$ and each

$$(x, u) \in X(A, B, 0, T\tau)$$

which satisfies for each $S \in [0, T\tau - L_1\tau]$,

$$I^g(S, S + L_1\tau, x, u) \leq \sigma(g, x(S), x(S + L_1\tau), S, S + L_1\tau) + \delta$$

and satisfies

$$|x(0) - x_f(0)| \leq \delta, \ |x(T\tau) - x_f(0)| \leq \delta,$$

$$I^g(0, T\tau, x, u) \leq \sigma(g, x(0), x(T\tau), 0, T\tau) + M_1$$

we have for all integers $i = 0, \dots, T - 1$,

$$|x(i\tau + t) - x_f(t)| \leq \varepsilon \text{ for all } t \in [0, \tau].$$

By Theorem 18, there exist an integer $L_2 \geq 1$ and a neighborhood $\mathscr{U} \subset \mathscr{U}_1$ of f in $\mathscr{M}$ such that the following property holds:

(P12) for each $T > L_2\tau$, each $g \in \mathscr{U}$, each $h \in \mathfrak{A}_n$ which satisfies

$$h(z) \leq M_2 \text{ for all } z \in R^n \text{ satisfying } |z| \leq \tilde{M}, \tag{9.1}$$

each $\xi \in \mathfrak{A}_{2n}$ which satisfy

$$\xi(z) \leq M_2 \text{ for all } z = (z_1, z_2) \in R^n \times R^n \text{ satisfying } |z_i| \leq \tilde{M},\ i = 1, 2 \tag{9.2}$$

and each $(x, u) \in X(A, B, 0, T)$ which satisfies at least one of the following conditions below

$$\text{(i) } |x(0)| \leq M_0,$$

$$I^g(0, T, x, u) + h(x(T)) \leq \sigma(g, h, x(0), 0, T) + M_1;$$

$$\text{(ii) } I^g(0, T, x, u) + \xi(x(0), x(T)) \leq \sigma(g, \xi, 0, T) + M_1$$

the following inequality holds:

$$\mathrm{Card}(\{i \in \{0, \ldots, \lfloor \tau^{-1}T \rfloor - 1\} :$$

$$\max\{|x(i\tau + t) - x_f(t)| : t \in [0, \tau]\} > \delta\}) \leq L_2.$$

Choose an integer

$$L \geq 4L_1 + 4L_2. \tag{9.3}$$

Assume that

$$T > L\tau,\ g \in \mathscr{U},\ h \in \mathfrak{A}_n,\ \xi \in \mathfrak{A}_{2n}, \tag{9.4}$$

Equations (9.1) and (9.2) hold and that $(x, u) \in X(A, B, 0, T)$ satisfies for each $S \in [0, T - L\tau]$,

$$I^g(S, S + L\tau, x, u) \leq \sigma(g, x(S), x(S + L\tau), S, S + L\tau) + \delta \tag{9.5}$$

and satisfies at least one of the conditions (i) and (ii). By (9.1)–(9.4), conditions (i) and (ii) and property (P12), there exist integers

$$p_1 \in [0, L_2],\ p_2 \in [\lfloor \tau^{-1}T \rfloor - L_2, \lfloor \tau^{-1}T \rfloor] \tag{9.6}$$

such that

$$|x(p_i\tau) - x_f(0)| \leq \delta,\ i = 1, 2. \tag{9.7}$$

If $|x(0) - x_f(0)| \le \delta$, then we set $p_1 = 0$ and if $|x(\lfloor \tau^{-1} T \rfloor \tau) - x_f(0)| \le \delta$, then we set $p_2 = \lfloor \tau^{-1} T \rfloor$. By (9.3)–(9.5), (9.7) and property (P11), for all integers $i = p_1, \dots, p_2 - 1$,

$$|x(i\tau + t) - x_f(t)| \le \varepsilon \text{ for all } t \in [0, \tau].$$

Theorem 19 is proved.

10 Auxiliary Results for Theorem 20

Lemma 1 *Let $h \in \mathfrak{A}_n$, $S_0 \ge 1$ be an integer, $\varepsilon \in (0, 1)$. Then there exists $\delta \in (0, \varepsilon)$ such that for each $(x, u) \in X(A, B, 0, S_0\tau)$ which satisfies*

$$(\pi^f + h)(x(0)) \le \inf(\pi^f + h) + \delta,$$

$$I^f(0, S_0\tau, x, u) - S_0\tau\mu(f) - \pi^f(x(0)) + \pi^f(x(S_0\tau)) \le \delta$$

there exists an (f, A, B)-overtaking optimal pair $(x_*, u_*) \in X(A, B, 0, \infty)$ such that

$$(\pi^f + h)(x_*(0)) = \inf(\pi^f + h),$$

$$|x(t) - x_*(t)| \le \varepsilon \textit{ for all } t \in [0, S_0\tau].$$

Proof Assume that the lemma does not hold. Then there is a sequence $\{\delta_k\}_{k=1}^{\infty} \subset (0, 1]$ and a sequence $\{(x_k, u_k)\}_{k=1}^{\infty} \subset X(A, B, 0, S_0\tau)$ such that

$$\lim_{k \to \infty} \delta_k = 0 \tag{10.1}$$

and that for all integers $k \ge 1$,

$$(\pi^f + h)(x_k(0)) \le \inf(\pi^f + h) + \delta_k, \tag{10.2}$$

$$I^f(0, S_0\tau, x_k, u_k) - S_0\tau\mu(f) - \pi^f(x_k(0)) + \pi^f(x_k(S_0\tau)) \le \delta_k, \tag{10.3}$$

and that the following property holds:

(i) for each (f, A, B)-overtaking optimal pair $(y, v) \in X(A, B, 0, \infty)$ satisfying

$$(\pi^f + h)(y(0)) = \inf(\pi^f + h)$$

we have

$$\sup\{|x_k(t) - y(t)| : t \in [0, S_0\tau]\} > \varepsilon.$$

In view of (10.2), (10.3) and the boundedness from below of the functions π^f, h, the sequences $\{\pi^f(x_k(0))\}_{k=1}^{\infty}$, $\{h(x_k(0))\}_{k=1}^{\infty}$, $\{I^f(0, S_0\tau, x_k, u_k)\}_{k=1}^{\infty}$ are bounded. By Proposition 17, extracting a subsequence and re-indexing if necessary, we may assume without loss of generality that there exists

$$(x, u) \in X(A, B, 0, S_0\tau)$$

such that

$$x_k(t) \to x(t) \text{ as } k \to \infty \text{ uniformly on } [0, S_0\tau], \tag{10.4}$$

$$I^f(0, S_0\tau, x, u) \le \liminf_{k\to\infty} I^f(0, S_0\tau, x_k, u_k), \tag{10.5}$$

$$u_k \to u \text{ as } k \to \infty \text{ weakly in } L^1(R^m; (0, S_0\tau)). \tag{10.6}$$

It follows from (10.1), (10.2), (10.4), the continuity of π^f and lower semicontinuity of h that

$$\pi^f(x(0)) = \lim_{k\to\infty} \pi^f(x_k(0)),\ \ h(x(0)) \le \liminf_{k\to\infty} h(x_k(0)),$$

$$(\pi^f + h)(x(0)) \le \liminf_{k\to\infty} (\pi^f + h)(x_k(0)) = \inf(\pi^f + h). \tag{10.7}$$

In view of (10.2) and (10.7),

$$h(x(0)) = \lim_{k\to\infty} h(x_k(0)). \tag{10.8}$$

By (10.4) and the continuity of π^f,

$$\pi^f(x(S_0\tau)) = \lim_{k\to\infty} \pi^f(x_k(S_0\tau)). \tag{10.9}$$

It follows from (10.1), (10.3), (10.5), (10.7) and (10.9) that

$$I^f(0, S_0\tau, x, u) - S_0\tau\mu(f) - \pi^f(x(0)) + \pi^f(x(S_0\tau))$$

$$\le \liminf_{k\to\infty}[I^f(0, S_0\tau, x_k, u_k) - S_0\tau\mu(f)] - \lim_{k\to\infty} \pi^f(x_k(0)) + \lim_{k\to\infty} \pi^f(x_k(S_0\tau))$$

$$= \liminf_{k\to\infty}[I^f(0, S_0\tau, x_k, u_k) - S_0\tau\mu(f) - \pi^f(x_k(0)) + \pi^f(x_k(S_0\tau))]$$

$$\le \lim_{k\to\infty} \delta_k = 0.$$

In view of the inequality above and Proposition 2,

$$I^f(0, S_0\tau, x, u) - S_0\tau\mu(f) - \pi^f(x(0)) + \pi^f(x(S_0\tau)) = 0. \tag{10.10}$$

Theorem 2 implies that there exists an (f, A, B)-overtaking optimal pair $(\tilde{x}, \tilde{u}) \in X(A, B, 0, \infty)$ such that $\tilde{x}(0) = x(S_0\tau)$. For all $t > S_0\tau$ set

$$x(t) = \tilde{x}(t - S_0\tau),\ u(t) = \tilde{u}(t - S_0\tau). \tag{10.11}$$

It is not difficult to see that the pair $(x, u) \in X(A, B, 0, \infty)$ is an (f, A, B)-good pair. By (10.10), (10.11) and Propositions 2 and 7,

$$I^f(0, S\tau, x, u) - S\tau\mu(f) - \pi^f(x(0)) + \pi^f(x(S\tau)) = 0 \text{ for all integers } S \geq 1.$$

Combined with Proposition 9 and (10.7) this implies that (x, u) is an (f, A, B)-overtaking optimal pair satisfying $(\pi^f + h)(x(0)) = \inf(\pi^f + h)$. By (10.4), for all sufficiently large natural numbers k,

$$|x_k(t) - x(t)| \leq \varepsilon/2 \text{ for all } t \in [0, S_0\tau].$$

This contradicts condition (i). The contradiction we have reached proves Lemma 1.

Note that Lemma 1 can also be applied for the triplet $(\bar{f}, -A, -B)$.

Lemma 2 *Let $h \in \mathfrak{A}_n$, $\widehat{S}_2 \geq \widehat{S}_1 \geq 1$ be integers, $\varepsilon \in (0, 1]$ and let $(\bar{x}_*, \bar{u}_*) \in X(-A, -B, 0, \infty)$ be an $(\bar{f}, -A, -B)$-overtaking optimal pair. Then there exist a neighborhood $\mathcal{U}$ of f in $\mathcal{M}$ and a neighborhood $\mathcal{V}$ of h in $\mathfrak{A}_n$ such that for each pair of integers $S_2 > S_1 \geq 0$ satisfying $S_2 - S_1 \in [\widehat{S}_1, \widehat{S}_2]$, each $g \in \mathcal{U}$, each $\xi \in \mathcal{V}$, $(y, v) \in X(A, B, S_1\tau, S_2\tau)$ such that*

$$y(t) = \bar{x}_*(S_2\tau - t),\ v(t) = \bar{u}_*(S_2\tau - t),\ t \in [S_1\tau, S_2\tau] \tag{10.12}$$

and each $(x, u) \in X(A, B, S_1\tau, S_2\tau)$ satisfying

$$I^g(S_1\tau, S_2\tau, x, u) + \xi(x(S_2\tau)) \leq I^g(S_1\tau, S_2\tau, y, v) + \xi(y(S_2\tau)) + \varepsilon \tag{10.13}$$

the following inequality holds:

$$I^f(S_1\tau, S_2\tau, x, u) + h(x(S_2\tau)) \leq I^f(S_1\tau, S_2\tau, y, v) + h(y(S_2\tau)) + 2\varepsilon$$

$$= I^{\bar{f}}(0, (S_2 - S_1)\tau, \bar{x}_*, \bar{u}_*) + h(\bar{x}_*(0)) + 2\varepsilon.$$

Proof Since $(\bar{x}_*, \bar{u}_*)$ is an $(\bar{f}, -A, -B)$-overtaking optimal pair it follows from Theorem 1 that

$$\sup\{|\bar{x}_*(t)| : t \in [0, \infty)\} < \infty.$$

Choose

$$M_0 > \sup\{|\bar{x}_*(t)| : t \in [0, \infty)\}. \tag{10.14}$$

Since $h \in \mathfrak{A}_n$ there exists $M_1 > 0$ such that

$$|h(z)| \le M_1 \text{ for all } z \in R^n \text{ satisfying } |z| \le M_0. \tag{10.15}$$

Since the function $\pi^{\bar{f}}$ is continuous there exists $M_2 > 0$ such that

$$|\pi^{\bar{f}}(z)| \le M_2 \text{ for all } z \in R^n \text{ satisfying } |z| \le M_0. \tag{10.16}$$

There exists a neighborhood $\mathscr{V}_0$ of h in $\mathfrak{A}_n$ such that for each $\xi \in \mathscr{V}_0$,

$$|\xi(z) - h(z)| \le \varepsilon/16 \text{ for all } z \in R^n \text{ satisfying } |z| \le M_0. \tag{10.17}$$

By Proposition 16, there exists a neighborhood $\mathscr{U}$ of f in $\mathscr{M}$ such that the following property holds:

(P13) for each $g \in \mathscr{U}$, each pair of integers $S_2 > S_1 \ge 0$ satisfying $S_2 - S_1 \le \widehat{S}_2$ and each $(x, u) \in X(A, B, S_1\tau, S_2\tau)$ satisfying

$$\min\{I^f(S_1\tau, S_2\tau, x, u), I^g(S_1\tau, S_2\tau, x, u)\} \le |\mu(f)|\widehat{S}_2\tau + 2M_1 + 2M_2 + a_1 + 4$$

we have $|I^f(S_1\tau, S_2\tau, x, u) - I^g(S_1\tau, S_2\tau, x, u)| \le \varepsilon/16$.

By Proposition 15, there exists $\Delta_0 > 0$ such that the following property holds:

(P14) for each $g \in \mathscr{M}$, each pair of integers $S_2 > S_1 \ge 0$ satisfying $S_2 - S_1 \le \widehat{S}_2$ and each $(x, u) \in X(A, B, S_1\tau, S_2\tau)$ satisfying

$$I^g(S_1\tau, S_2\tau, x, u) \le |\mu(f)|\widehat{S}_2\tau + 2M_1 + 2M_2 + a_1 + 4$$

we have $|x(t)| \le \Delta_0$ for all $t \in [S_1\tau, S_2\tau]$.

There exists a neighborhood $\mathscr{V} \subset \mathscr{V}_0$ of h in $\mathfrak{A}_n$ such that for each $\xi \in V$,

$$|\xi(z) - h(z)| \le \varepsilon/16 \text{ for all } z \in R^n \text{ satisfying } |z| \le M_0 + \Delta_0. \tag{10.18}$$

Assume that integers $S_2 > S_1 \ge 0$ satisfy

$$S_2 - S_1 \in [\widehat{S}_1, \widehat{S}_2],\ g \in \mathscr{U},\ \xi \in \mathscr{V}, \tag{10.19}$$

$(y, v) \in X(-A, -B, S_1\tau, S_2\tau)$ satisfies (10.12) and that

$$(x, u) \in X(A, B, S_1\tau, S_2\tau)$$

satisfies (10.13). In view of (3.6) and (10.12),

$$I^f(S_1\tau, S_2\tau, y, v) = I^{\bar{f}}(0, (S_2 - S_2)\tau, \bar{x}_*, \bar{u}_*). \tag{10.20}$$

Since $(\bar{x}_*, \bar{u}_*)$ is an $(\bar{f}, -A, -B)$-overtaking optimal pair Proposition 3 implies that

$$I^{\bar{f}}(0, (S_2 - S_1)\tau, \bar{x}_*, \bar{u}_*) = \mu(f)(S_2\tau - S_1\tau) + \pi^{\bar{f}}(\bar{x}_*(0)) - \pi^{\bar{f}}(\bar{x}_*(S_2\tau - S_1\tau)). \tag{10.21}$$

Set

$$S = S_2 - S_1. \tag{10.22}$$

In view of (10.14), (10.16) and (10.20)–(10.22),

$$I^f(S_1\tau, S_2\tau, y, v) \leq \mu(f)S\tau + 2M_2. \tag{10.23}$$

Property (P13), (10.19) and (10.23) imply that

$$I^g(S_1\tau, S_2\tau, y, v) \leq I^f(S_1\tau, S_2\tau, y, v) + \varepsilon/16. \tag{10.24}$$

By (10.12), (10.14) and (10.17),

$$|\xi(y(S_2\tau)) - h(y(S_2\tau))| = |\xi(\bar{x}_*(0)) - h(\bar{x}_*(0))| \leq \varepsilon/16. \tag{10.25}$$

In view of (10.13), (10.24) and (10.25),

$$I^g(S_1\tau, S_2\tau, x, u) + \xi(x(S_2\tau)) \leq I^f(S_1\tau, S_2\tau, y, v) + h(y(S_2\tau)) + \varepsilon/8 + \varepsilon. \tag{10.26}$$

It follows from (4.3), (10.12), (10.14), (10.15), (10.23) and (10.26) that

$$\begin{aligned} I^g(S_1\tau, S_2\tau, x, u) &\leq I^f(S_1\tau, S_2\tau, y, v) + h(y(S_2\tau)) + \varepsilon/8 + \varepsilon + a_1 \\ &\leq \mu(f)S\tau + 2M_2 + h(\bar{x}_*(0)) + a_1 + \varepsilon + \varepsilon/8. \end{aligned} \tag{10.27}$$

Property (P13), (10.19), (10.22) and (10.27) imply that

$$|I^f(S_1\tau, S_2\tau, x, u) - I^g(S_1\tau, S_2\tau, x, u)| \leq \varepsilon/16. \tag{10.28}$$

By property (P14), (10.14), (10.15), (10.19) and (10.27),

$$|x(t)| \leq \Delta_0, \ t \in [S_1\tau, S_2\tau]. \tag{10.29}$$

In view of (10.18), (10.19) and (10.29),

$$|\xi(x(S_2\tau)) - h(x(S_2\tau))| \leq \varepsilon/16. \tag{10.30}$$

It follows from (10.12), (10.20), (10.26), (10.28) and (10.30) that

$$I^f(S_1\tau, S_2\tau, x, u) + h(x(S_2\tau)) \le I^g(S_1\tau, S_2\tau, x, u) + \xi(x(S_2\tau)) + \varepsilon/8$$

$$\le I^f(S_1\tau, S_2\tau, y, v) + h(y(S_2\tau)) + \varepsilon/8 + \varepsilon + \varepsilon/8$$

$$= I^{\bar{f}}(0, (S_2 - S_1)\tau, \bar{x}_*, \bar{u}_*) + h(\bar{x}_*(0)) + \varepsilon + \varepsilon/4.$$

Lemma 2 is proved.

11 Proof of Theorem 20

By Lemma 1 applied to the triplet $(\bar{f}, -A - B)$ there exist $\delta_1 \in (0, \varepsilon/4)$ such that the following property holds:

(P15) for each $(x, u) \in X(-A, -B, 0, L_0\tau)$ which satisfies

$$(\pi^{\bar{f}} + h)(x(0)) \le \inf(\pi^{\bar{f}} + h) + 4\delta_1,$$

$$I^{\bar{f}}(0, L_0\tau, x, u) - L_0\tau\mu(f) - \pi^{\bar{f}}(x(0)) + \pi^{\bar{f}}(x(L_0\tau)) \le 4\delta_1$$

there exists an $(\bar{f}, -A, -B)$-overtaking optimal pair

$$(\widehat{x}, \widehat{u}) \in X(-A-, B, 0, \infty)$$

such that

$$(\pi^{\bar{f}} + h)(\widehat{x}(0)) = \inf(\pi^{\bar{f}} + h),$$

$$|x(t) - \widehat{x}(t)| \le \varepsilon \text{ for all } t \in [0, L_0\tau].$$

In view of the continuity of $\pi^{\bar{f}}$, Proposition 4 and (5.4), there exists $\delta_2 \in (0, \delta_1)$ such that for each $z \in R^n$ satisfying $|z - x_f(0)| \le 2\delta_2$,

$$|\pi^{\bar{f}}(z)| = |\pi^{\bar{f}}(z) - \pi^{\bar{f}}(x_f(0))| \le \delta_1/8; \tag{11.1}$$

for each $y, z \in R^n$ satisfying $|y - x_f(0)| \le 2\delta_2$, $|z - x_f(0)| \le 2\delta_2$,

$$|v(y, z) - \tau\mu(f)| \le \delta_1/8. \tag{11.2}$$

By Theorem 17, there exist an integer $l_0 \ge 1$, $\delta_3 \in (0, \delta_2/8)$, a neighborhood $\mathscr{U}_1$ of f in $\mathscr{M}$ and a neighborhood $\mathscr{V}_1$ of h in $\mathfrak{A}_n$ such that the following property holds:

(P16) for each integer $T > 2l_0$, each $g \in \mathcal{U}_1$, each $\xi \in \mathcal{V}_1$ and each

$$(x, u) \in X(A, B, 0, T\tau)$$

such that

$$|x(0)| \leq M, \ I^g(0, T\tau, x, u) + \xi(x(T\tau)) \leq \sigma(g, \xi, x(0), 0, T\tau) + \delta_3$$

we have

$$|x(i\tau) - x_f(0)| \leq \delta_2 \text{ for all } i = l_0, \ldots, T - l_0.$$

By Theorem 2 and Proposition 12, there exists $(\bar{f}, -A, -B)$-overtaking optimal pair

$$(\bar{x}_*, \bar{u}_*) \in X(-A-, B, 0, \infty) \tag{11.3}$$

such that

$$(\pi^{\bar{f}} + h)(\bar{x}_*(0)) = \inf(\pi^{\bar{f}} + h). \tag{11.4}$$

Since the pair $(\bar{x}_*, \bar{u}_*) \in X(-A, -B, 0, \infty)$ is $(\bar{f}, -A, -B)$-good it follows from Theorems 1 and 8 that there exists an integer $l_1 \geq 1$ such that

$$|\bar{x}^*(i\tau) - x_f(0)| \leq \delta_2 \text{ for all integers } i \geq l_1. \tag{11.5}$$

By Proposition 16, there exists a neighborhood $\mathcal{U}_2 \subset \mathcal{U}_1$ of f in $\mathcal{M}$ such that the following property holds:

(P17) for each $g \in \mathcal{U}_2$ and each $(x, u) \in X(A, B, 0, \tau)$ satisfying

$$\min\{I^f(0, \tau, x, u), I^g(0, \tau, x, u)\} \leq |\mu(f)|\tau + 2$$

we have $|I^f(0, \tau, x, u) - I^g(0, \tau, x, u)| \leq \delta_3/8$.

By Lemma 2, there exist a neighborhood $\mathcal{U} \subset \mathcal{U}_2$ of f in $\mathcal{M}$ and a neighborhood $\mathcal{V} \subset \mathcal{V}_1$ of h in $\mathfrak{A}_n$ such that the following property holds:

(P18) for each pair of integers $S_2 > S_1 \geq 0$ satisfying $S_2 - S_1 \in [1, 2L_0 + 2l_0 + l_1 + 4]$, each $g \in \mathcal{U}$, each $\xi \in \mathcal{V}$, $(y, v) \in X(A, B, S_1\tau, S_2\tau)$ such that

$$y(t) = \bar{x}_*(S_2\tau - t), \ v(t) = \bar{u}_*(S_2\tau - t), \ t \in [S_1\tau, S_2\tau]$$

and each $(x, u) \in X(A, B, S_1\tau, S_2\tau)$ satisfying

$$I^g(S_1\tau, S_2\tau, x, u) + \xi(x(S_2\tau)) \leq I^g(S_1\tau, S_2\tau, y, v) + \xi(y(S_2\tau)) + \delta_1$$

we have

$$I^f(S_1\tau, S_2\tau, x, u) + h(x(S_2\tau)) \leq I^{\bar{f}}(0, (S_2 - S_1)\tau, \bar{x}_*, \bar{u}_*) + h(\bar{x}_*(0)) + 2\delta_1.$$

Choose $\delta > 0$ and an integer L_1 such that

$$\delta \leq \delta_3/4, \tag{11.6}$$

$$L_1 > 2L_0 + 2l_0 + 2l_1 + 4. \tag{11.7}$$

Assume that an integer

$$T \geq L_1,\ g \in \mathscr{U},\ \xi \in \mathscr{V},\ (x, u) \in X(A, B, 0, T\tau), \tag{11.8}$$

$$|x(0)| \leq M, \tag{11.9}$$

$$I^g(0, T\tau, x, u) + \xi(x(T\tau)) \leq \sigma(g, \xi, x(0), 0, T\tau) + \delta. \tag{11.10}$$

Property (P16) and (11.6)–(11.10) imply that

$$|x(i\tau) - x_f(0)| \leq \delta_2 \text{ for all } i \in \{l_0, \dots, T - l_0\}. \tag{11.11}$$

In view of (11.7) and (11.8),

$$[T - l_0 - l_1 - L_0 - 4, T - l_0 - l_1 - L_0] \subset [l_0, T - l_0 - l_1 - L_0]. \tag{11.12}$$

By (11.11) and (11.12),

$$|x(i\tau) - x_f(0)| \leq \delta_2 \text{ for all } i \in \{T - l_0 - l_1 - L_0 - 4, \dots, T - l_0 - l_1 - L_0\}. \tag{11.13}$$

Proposition 18 implies that there exists $(x_1, u_1) \in X(A, B, 0, T\tau)$ such that

$$x_1(t) = x(t),\ u_1(t) = u(t),\ t \in [0, \tau(T - l_0 - l_1 - L_0 - 4)],$$

$$x_1(t) = \bar{x}_*(T\tau - t),\ u_1(t) = \bar{u}_*(T\tau - t),\ t \in [\tau(T - l_0 - l_1 - L_0 - 3), \tau T],$$

$$I^f(\tau(T - l_0 - l_1 - L_0 - 4), \tau(T - l_0 - l_1 - L_0 - 3), x_1, u_1)$$

$$= v(x(\tau(T - l_0 - l_1 - L_0 - 4)), \bar{x}_*(\tau(l_0 + l_1 + L_0 + 3))). \tag{11.14}$$

By (11.10) and (11.14),

$$-\delta \leq I^g(0, T\tau, x_1, u_1) + \xi(x_1(T\tau)) - (I^g(0, T\tau, x, u) + \xi(x(T\tau)))$$

$$= I^g(\tau(T - l_0 - l_1 - L_0 - 4), \tau(T - l_0 - l_1 - L_0 - 3), x_1, u_1)$$

$$+ I^g(\tau(T - l_0 - l_1 - L_0 - 3), \tau T, x_1, u_1) + \xi(x_1(T\tau))$$

$$-I^g(\tau(T-l_0-l_1-L_0-4),\tau(T-l_0-l_1-L_0-3),x,u)$$

$$-I^g(\tau(T-l_0-l_1-L_0-3),\tau T,x,u)-\xi(x(T\tau)). \tag{11.15}$$

We will estimate

$$I^g(\tau(T-l_0-l_1-L_0-4),\tau(T-l_0-l_1-L_0-3),x_1,u_1)$$

$$-I^g(\tau(T-l_0-l_1-L_0-4),\tau(T-l_0-l_1-L_0-3),x,u).$$

In view of (11.2), (11.5), (11.13) and (11.14),

$$I^f(\tau(T-l_0-l_1-L_0-4),\tau(T-l_0-l_1-L_0-3),x_1,u_1)$$

$$=v(x(\tau(T-l_0-l_1-L_0-4)),\bar{x}_*(\tau(l_0+l_1+L_0+3)))\le\tau\mu(f)+\delta_1/8.$$

Combined with (11.8) and property (P17) this implies that

$$I^g(\tau(T-l_0-l_1-L_0-4),\tau(T-l_0-l_1-L_0-3),x_1,u_1)$$

$$\le I^f(\tau(T-l_0-l_1-L_0-4),\tau(T-l_0-l_1-L_0-3),x_1,u_1)+\delta_3/8$$

$$\le\tau\mu(f)+\delta_1/8+\delta_3/8. \tag{11.16}$$

It follows from (11.2) and (11.13) that

$$I^f(\tau(T-l_0-l_1-L_0-4),\tau(T-l_0-l_1-L_0-3),x,u)$$

$$\ge v(x(\tau(T-l_0-l_1-L_0-4)),x(\tau(T-l_0-l_1-L_0-3)))\ge\tau\mu(f)-\delta_1/8. \tag{11.17}$$

We claim that

$$I^g(\tau(T-l_0-l_1-L_0-4),\tau(T-l_0-l_1-L_0-3),x,u)\ge\tau\mu(f)-\delta_1/2. \tag{11.18}$$

Assume the contrary. Then

$$I^g(\tau(T-l_0-l_1-L_0-4),\tau(T-l_0-l_1-L_0-3),x,u)$$

$$<\tau\mu(f)-\delta_1/2. \tag{11.19}$$

By property (P17), (11.8) and (11.19),

$$I^f(\tau(T - l_0 - l_1 - L_0 - 4), \tau(T - l_0 - l_1 - L_0 - 3), x, u)$$

$$\leq I^g(\tau(T - l_0 - l_1 - L_0 - 4), \tau(T - l_0 - l_1 - L_0 - 3), x, u) + \delta_3/8$$

$$< \tau\mu(f) - \delta_1/2 + \delta_3/8 < \tau\mu(f) - \delta_1/4.$$

This contradicts (11.17). The contradiction we have reached proves (11.18). It follows from (11.16) and (11.18) that

$$I^g(\tau(T - l_0 - l_1 - L_0 - 4), \tau(T - l_0 - l_1 - L_0 - 3), x_1, u_1)$$

$$- I^g(\tau(T - l_0 - l_1 - L_0 - 4), \tau(T - l_0 - l_1 - L_0 - 3), x, u) \leq \delta_1/8 + \delta_3/8 + \delta_1/2. \tag{11.20}$$

By (11.15),

$$I^g(\tau(T - l_0 - l_1 - L_0 - 3), \tau T, x_1, u_1) + \xi(x_1(T\tau))$$

$$-I^g(\tau(T - l_0 - l_1 - L_0 - 3), \tau T, x, u) + \xi(x(T\tau))$$

$$\geq -\delta - I^g(\tau(T - l_0 - l_1 - L_0 - 4), \tau(T - l_0 - l_1 - L_0 - 3), x_1, u_1)$$

$$+I^g(\tau(T - l_0 - l_1 - L_0 - 4), \tau(T - l_0 - l_1 - L_0 - 3), x, u)$$

$$\geq -\delta - 5\delta_1/8 - \delta_3/8 \geq -\delta_1. \tag{11.21}$$

By (11.1), (11.5) and the choice of δ_2,

$$|\pi^{\bar{f}}(\bar{x}_*(\tau(l_0 + l_1 + L_0 + 3)))| \leq \delta_1/8. \tag{11.22}$$

In view of (11.21),

$$I^g(\tau(T - l_0 - l_1 - L_0 - 3), \tau T, x, u) + \xi(x(T\tau))$$

$$\leq I^g(\tau(T - l_0 - l_1 - L_0 - 3), \tau T, x_1, u_1) + \xi(x_1(T\tau)) + \delta_1. \tag{11.23}$$

By (11.8), (11.23) and property (P18) applied with $(y, v) = (x_1, u_1)$,

$$I^f(\tau(T - l_0 - l_1 - L_0 - 3), \tau T, x, u) + h(x(T\tau))$$

$$\leq I^{\bar{f}}(0, (L_0 + l_0 + l_1 + 3)\tau, \bar{x}_*, \bar{u}_*) + h(\bar{x}_*(0)) + 2\delta_1. \tag{11.24}$$

Proposition 3, (11.22) and (11.24) imply that

$$I^f(\tau(T-l_0-l_1-L_0-3),\tau T,x,u)+h(x(T\tau))$$
$$\le \mu(f)\tau(l_0+l_1+L_0+3)+\pi^{\bar{f}}(\bar{x}_*(0))$$
$$-\pi^{\bar{f}}(\bar{x}_*((l_0+l_1+L_0+3)\tau))+h(\bar{x}_*(0))+2\delta_1$$
$$\le \mu(f)\tau(l_0+l_1+L_0+3)+\pi^{\bar{f}}(\bar{x}_*(0))+h(\bar{x}_*(0))+2\delta_1+\delta_1/8. \tag{11.25}$$

Set

$$\tilde{x}(t)=x(T\tau-t),\ \tilde{u}(t)=u(T\tau-t),\ t\in[0,\tau(L_0+l_0+l_1+3)]. \tag{11.26}$$

In view of (3.6), (11.25) and (11.26),

$$I^{\bar{f}}(0,\tau(l_0+l_1+L_0+3),\tilde{x},\tilde{u})+h(\tilde{x}(0))$$
$$=I^f(\tau(T-l_0-l_1-L_0-3),\tau T,x,u)+h(x(\tau T))$$
$$\le \mu(f)\tau(l_0+l_1+L_0+3)+\pi^{\bar{f}}(\bar{x}_*(0))+h(\bar{x}_*(0))+2\delta_1+\delta_1/8. \tag{11.27}$$

It follows from (11.13) and (11.26) that

$$|\tilde{x}(\tau(l_0+l_1+L_0+3))-x_f(0)|\le\delta_2. \tag{11.28}$$

By (11.1) and (11.28),

$$|\pi^{\bar{f}}(\tilde{x}(\tau(l_0+l_1+L_0+3)))|\le\delta_1/8. \tag{11.29}$$

By (11.27), (11.29) and Proposition 2,

$$(\pi^{\bar{f}}+h)(\tilde{x}(0))-(\pi^{\bar{f}}+h)(\bar{x}_*(0))$$
$$+I^{\bar{f}}(0,L_0\tau,\tilde{x},\tilde{u})-L_0\tau\mu(f)-\pi^{\bar{f}}(\tilde{x}(0))+\pi^{\bar{f}}(\tilde{x}(L_0\tau))$$
$$\le(\pi^{\bar{f}}+h)(\tilde{x}(0))-(\pi^{\bar{f}}+h)(\bar{x}_*(0))$$
$$+I^{\bar{f}}(0,\tau(l_0+l_1+L_0+3),\tilde{x},\tilde{u})$$
$$-\mu(f)\tau(l_0+l_1+L_0+3)-\pi^{\bar{f}}(\tilde{x}(0))+\pi^{\bar{f}}(\tilde{x}(\tau(l_0+l_1+L_0+3)))$$
$$\le\pi^{\bar{f}}(\tilde{x}(0))-\pi^{\bar{f}}(\bar{x}_*(0))-h(\bar{x}_*(0))$$

$$+\mu(f)\tau(l_0+l_1+L_0+3)+\pi^{\bar f}(\bar x_*(0))+h(\bar x_*(0))+2\delta_1+\delta_1/8$$

$$-\mu(f)\tau(l_0+l_1+L_0+3)-\pi^{\bar f}(\tilde x(0))+\pi^{\bar f}(\tilde x((l_0+l_1+L_0+3)\tau))$$

$$\le 2\delta_1+\delta_1/8+\delta_1/8. \tag{11.30}$$

By Proposition 2, (11.4) and (11.30),

$$(\pi^{\bar f}+h)(\tilde x(0))\le(\pi^{\bar f}+h)(\bar x_*(0))+3\delta_1\le\inf(\pi^{\bar f}+h)+3\delta_1,$$

$$I^{\bar f}(0,L_0\tau,\tilde x,\tilde u)-L_0\tau\mu(f)+\pi^{\bar f}(\tilde x(0))+\pi^{\bar f}(\tilde x(L_0\tau))\le 3\delta_1.$$

It follows from the inequalities above, (11.26) and property (P15) that there exists an $(\bar f,-A,-B)$-overtaking optimal pair $(\widehat{x},\widehat{u})\in X(-A-,B,0,\infty)$ such that

$$(\pi^{\bar f}+h)(\widehat{x}(0))=\inf(\pi^{\bar f}+h)$$

and for all $t\in[0,L_0\tau]$,

$$\varepsilon\ge|\tilde x(t)-\widehat{x}(t)|=|x(T\tau-t)-\widehat{x}(t)|.$$

Theorem 20 is proved.

12 Auxiliary Results for Theorem 21

Recall (see (4.4)) that for each $h\in\mathfrak{A}_{2n}$

$$\psi_h(z_1,z_2)=\pi^f(z_1)+\pi^{\bar f}(z_2)+h(z_1,z_2)\text{ for all }(z_1,z_2)\in R^n\times R^n. \tag{12.1}$$

Lemma 3 *Let $h\in\mathfrak{A}_{2n}$, $S_0\ge 1$ be an integer, $\varepsilon\in(0,1)$. Then there exists $\delta\in(0,\varepsilon)$ such that for each $(x_1,u_1)\in X(A,B,0,S_0\tau)$ and each*

$$(x_2,u_2)\in X(-A,-B,0,S_0\tau)$$

which satisfy

$$\psi_h(x_1(0),x_2(0))\le\inf(\psi_h)+\delta,$$

$$I^f(0,S_0\tau,x_1,u_1)-S_0\tau\mu(f)-\pi^f(x_1(0))+\pi^f(x_1(S_0\tau))\le\delta,$$

$$I^{\bar f}(0,S_0\tau,x_2,u_2)-S_0\tau\mu(f)-\pi^{\bar f}(x_2(0))+\pi^{\bar f}(x_2(S_0\tau))\le\delta$$

there exist an (f, A, B)*-overtaking optimal pair* $(x_1^*, u_1^*) \in X(A, B, 0, \infty)$ *and an* $(\bar{f}, -A, -B)$*-overtaking optimal pair* $(x_2^*, u_2^*) \in X(-A, -B, 0, \infty)$ *such that*

$$\psi_h(x_1^*(0), x_2^*(0)) = \inf(\psi_h)$$

and that for all $t \in [0, S_0\tau]$,

$$|x_1(t) - x_1^*(t)| \le \varepsilon, \ |x_2(t) - x_2^*(t)| \le \varepsilon.$$

Proof Assume that the lemma does not hold. Then there are a sequence $\{\delta_k\}_{k=1}^{\infty} \subset (0, 1]$ and sequences

$$\{(x_{k,1}, u_{k,1})\}_{k=1}^{\infty} \subset X(A, B, 0, S_0\tau), \ \{(x_{k,2}, u_{k,2})\}_{k=1}^{\infty} \subset X(-A, -B, 0, S_0\tau)$$

such that

$$\lim_{k\to\infty} \delta_k = 0, \tag{12.2}$$

for all integers $k \ge 1$,

$$\psi_h(x_{k,1}(0), x_{k,2}(0)) \le \inf(\psi_h) + \delta_k, \tag{12.3}$$

$$I^f(0, S_0\tau, x_{k,1}, u_{k,1}) - S_0\tau\mu(f) - \pi^f(x_{k,1}(0)) + \pi^f(x_{k,1}(S_0\tau)) \le \delta_k, \tag{12.4}$$

$$I^{\bar{f}}(0, S_0\tau, x_{k,2}, u_{k,2}) - S_0\tau\mu(f) - \pi^{\bar{f}}(x_{k,2}(0)) + \pi^{\bar{f}}(x_{k,2}(S_0\tau)) \le \delta_k \tag{12.5}$$

and that for each integer $k \ge 1$, each (f, A, B)-overtaking optimal pair $(\xi_1, \eta_1) \in X(A, B, 0, \infty)$ and each $(\bar{f}, -A, -B)$-overtaking optimal pair

$$(\xi_2, \eta_2) \in X(-A, -B, 0, \infty)$$

satisfying

$$\psi_h(\xi_1(0), \xi_2(0)) = \inf(\psi_h) \tag{12.6}$$

we have

$$\sup\{|\xi_1(t) - x_{k,1}(t)|, \ |\xi_2(t) - x_{k,2}(t)| : \ t \in [0, S_0\tau]\} > \varepsilon. \tag{12.7}$$

In view of (12.1)–(12.3) and the boundedness from below of the functions π^f, $\pi^{\bar{f}}$, h, the sequences $\{\pi^f(x_{k,1}(0))\}_{k=1}^{\infty}$, $\{\pi^{\bar{f}}(x_{k,2}(0))\}_{k=1}^{\infty}$ and

$$\{h(x_{k,1}(0), x_{k,2}(0))\}_{k=1}^{\infty}$$

are bounded. Together with (12.4), (12.5) and the boundedness from below of the functions $\pi^f, \pi^{\bar{f}}$ this implies that the sequences

$$\{I^f(0, S_0\tau, x_{k,1}, u_{k,1})\}_{k=1}^{\infty}, \ \{I^{\bar{f}}(0, S_0\tau, x_{k,2}, u_{k,2})\}_{k=1}^{\infty}$$

are bounded. By Proposition 17, extracting a subsequence and re-indexing if necessary, we may assume without loss of generality that there exists $(x_1, u_1) \in X(A, B, 0, S_0\tau)$ and $(x_2, u_2) \in X(-A, -B, 0, S_0\tau)$ such that for $i = 1, 2$,

$$x_{k,i}(t) \to x_i(t) \text{ as } k \to \infty \text{ uniformly on } [0, S_0\tau], \tag{12.8}$$

$$I^f(0, S_0\tau, x_1, u_1) \le \liminf_{k\to\infty} I^f(0, S_0\tau, x_{k,1}, u_{k,1}), \tag{12.9}$$

$$I^{\bar{f}}(0, S_0\tau, x_2, u_2) \le \liminf_{k\to\infty} I^{\bar{f}}(0, S_0\tau, x_{k,2}, u_{k,2}). \tag{12.10}$$

It follows from (12.2), (12.3), (12.8), the continuity of $\pi^f, \pi^{\bar{f}}$ and the lower semicontinuity of $h, \ \psi_h$ that

$$\pi^f(x_1(0)) = \lim_{k\to\infty} \pi^f(x_{k,1}(0)), \ \pi^{\bar{f}}(x_2(0)) = \lim_{k\to\infty} \pi^{\bar{f}}(x_{k,2}(0)), \tag{12.11}$$

$$h(x_1(0), x_2(0)) \le \liminf_{k\to\infty} h(x_{k,1}(0), x_{k,2}(0)), \tag{12.12}$$

$$\psi_h(x_1(0), x_2(0)) \le \liminf_{k\to\infty} \psi_h(x_{k,1}(0), x_{k,2}(0)) = \inf(\psi_h), \tag{12.13}$$

$$\pi^f(x_1(S_0\tau)) = \lim_{k\to\infty} \pi^f(x_{k,1}(S_0\tau)), \ \pi^{\bar{f}}(x_2(S_0\tau)) = \lim_{k\to\infty} \pi^{\bar{f}}(x_{k,2}(S_0\tau)). \tag{12.14}$$

In view of (12.1)–(12.3) and (12.11)–(12.13),

$$h(x_1(0), x_2(0)) = \lim_{k\to\infty} h(x_{k,1}(0), (x_{k,2}(0)). \tag{12.15}$$

It follows from (12.4), (12.5) and (12.9)–(12.11) that

$$I^f(0, S_0\tau, x_1, u_1) - S_0\tau\mu(f) - \pi^f(x_1(0)) + \pi^f(x_1(S_0\tau))$$

$$\le \liminf_{k\to\infty}[I^f(0, S_0\tau, x_{k,1}, u_{k,1}) - S_0\tau\mu(f)$$

$$- \pi^f(x_{k,1}(0)) + \pi^f(x_{k,1}(S_0\tau))] \le \lim_{k\to\infty} \delta_k = 0, \tag{12.16}$$

$$I^{\bar{f}}(0, S_0\tau, x_2, u_2) - S_0\tau\mu(f) - \pi^{\bar{f}}(x_2(0)) + \pi^{\bar{f}}(x_2(S_0\tau))$$

$$\leq \liminf_{k\to\infty}[I^{\bar{f}}(0, S_0\tau, x_{k,2}, u_{k,2}) - S_0\tau\mu(f)$$

$$- \pi^{\bar{f}}(x_{k,2}(0)) + \pi^{\bar{f}}(x_{k,2}(S_0\tau))] \leq \lim_{k\to\infty} \delta_k = 0. \tag{12.17}$$

In view of (12.16), (12.17) and Proposition 2,

$$I^f(0, S_0\tau, x_1, u_1) - S_0\tau\mu(f) - \pi^f(x_1(0)) + \pi^f(x_1(S_0\tau)) = 0, \tag{12.18}$$

$$I^{\bar{f}}(0, S_0\tau, x_2, u_2) - S_0\tau\mu(f) - \pi^{\bar{f}}(x_2(0)) + \pi^{\bar{f}}(x_2(S_0\tau)) = 0. \tag{12.19}$$

Theorem 2 implies that there is an (f, A, B)-overtaking optimal pair $(\tilde{x}_1, \tilde{u}_1) \in X(A, B, 0, \infty)$ such that

$$\tilde{x}_1(0) = x_1(S_0\tau) \tag{12.20}$$

and an $(\bar{f}, -A, -B)$-overtaking optimal pair $(\tilde{x}_2, \tilde{u}_2) \in X(-A, -B, 0, \infty)$ such that

$$\tilde{x}_2(0) = x_2(S_0\tau). \tag{12.21}$$

For all $t > S_0\tau$ and $i = 1, 2$ set

$$x_i(t) = \tilde{x}_i(t - S_0\tau),\ u_i(t) = \tilde{u}_i(t - S_0\tau). \tag{12.22}$$

It is not difficult to see that the pair $(x_1, u_1) \in X(A, B, 0, \infty)$ is an (f, A, B)-good pair and that the pair $(x_2, u_2) \in X(-A, -B, 0, \infty)$ is an $(\bar{f}, -A, -B)$-good pair. By (12.18), (12.19), (12.21), (12.22) and Propositions 2 and 3, for all integers $S > 0$,

$$I^f(0, S\tau, x_1, u_1) - S\tau\mu(f) - \pi^f(x_1(0)) + \pi^f(x_1(S\tau)) = 0,$$

$$I^{\bar{f}}(0, S\tau, x_2, u_2) - S\tau\mu(f) - \pi^{\bar{f}}(x_2(0)) + \pi^{\bar{f}}(x_2(S\tau)) = 0.$$

Combined with Proposition 9 this implies that $(x_1, u_1) \in X(A, B, 0, \infty)$ is an (f, A, B)-overtaking optimal pair and that the pair $(x_2, u_2) \in X(-A, -B, 0, \infty)$ is an $(\bar{f}, -A, -B)$-overtaking optimal. By (12.13),

$$\psi_h(x_1(0), x_2(0)) = \inf(\psi_h).$$

By (12.8), for all sufficiently large natural numbers k and $i = 1, 2$,

$$|x_{k,i}(t) - x_i(t)| \leq \varepsilon/2 \text{ for all } t \in [0, S_0\tau].$$

This contradicts (12.7). The contradiction we have reached proves Lemma 3.

Lemma 4 *Let* $h \in \mathfrak{A}_{2n}$, $\widehat{S}_2 \geq \widehat{S}_1 \geq 1$ *be integers,* $\varepsilon \in (0, 1)$ *and* $M > 0$. *Then there exist a neighborhood* $\mathcal{U}$ *of* f *in* $\mathcal{M}$ *and a neighborhood* $\mathcal{V}$ *of* h *in* $\mathfrak{A}_{2n}$ *such that for each* $g \in \mathcal{U}$, *each* $\xi \in \mathcal{V}$, *each pair of integers* $S_{i,2} > S_{i,1} \geq 0$, $i = 1, 2$ *satisfying*

$$S_{i,2} - S_{i,1} \in [\widehat{S}_1, \widehat{S}_2],\ i = 1, 2, \tag{12.23}$$

each $(x_1, u_1) \in X(A, B, S_{1,1}\tau, S_{1,2}\tau)$ *and each*

$$(x_2, u_2) \in X(A, B, S_{2,1}\tau, S_{2,2}\tau)$$

which satisfy

$$I^g(S_{1,1}\tau, S_{1,2}\tau, x_1, u_1) + I^g(S_{2,1}\tau, S_{2,2}\tau, x_2, u_2) + \xi(x_1(S_{1,1}\tau), x_2(S_{2,2}\tau)) \leq M \tag{12.24}$$

the following inequalities hold:

$$|I^g(S_{i,1}\tau, S_{i,2}\tau, x_i, u_i) - I^f(S_{i,1}\tau, S_{i,2}\tau, x_i, u_i)| \leq \varepsilon,\ i = 1, 2,$$

$$|h(x_1(S_{1,1}\tau), x_2(S_{2,2}\tau)) - \xi(x_1(S_{1,1}\tau), x_2(S_{2,2}\tau))| \leq \varepsilon.$$

Proof By Proposition 16, there exists a neighborhood $\mathcal{U}$ of f in $\mathcal{M}$ such that the following property holds:

(P19) for each $g \in \mathcal{U}$, each pair of integers $S_2 > S_1 \geq 0$ satisfying $S_2 - S_1 \leq \widehat{S}_2$ and each $(x, u) \in X(A, B, S_1\tau, S_2\tau)$ satisfying

$$\min\{I^f(S_1\tau, S_2\tau, x, u), I^g(S_1\tau, S_2\tau, x, u)\} \leq M + a_1 + a\tau S_2$$

we have $|I^f(S_1\tau, S_2\tau, x, u) - I^g(S_1\tau, S_2\tau, x, u)| \leq \varepsilon$.

By Proposition 15, there exists $\Delta_0 > 0$ such that the following property holds:

(P20) for each $g \in \mathcal{M}$, each pair of integers $S_2 > S_1 \geq 0$ satisfying $S_2 - S_1 \leq \widehat{S}_2$ and each $(x, u) \in X(A, B, S_1\tau, S_2\tau)$ satisfying

$$I^g(S_1\tau, S_2\tau, x, u) \leq M + a_1 + a\tau S_2$$

we have $|x(t)| \leq \Delta_0$ for all $t \in [S_1\tau, S_2\tau]$.

There exist a neighborhood $\mathcal{V}$ of h in $\mathfrak{A}_{2n}$ such that for each $\xi \in V$,

$$|\xi(z_1, z_2) - h(z_1, z_2)| \leq \varepsilon \text{ for all } z_1, z_2 \in R^n \text{ satisfying } |z_i| \leq \Delta_0,\ i = 1, 2. \tag{12.25}$$

Assume that

$$g \in \mathcal{U},\ \xi \in \mathcal{V}, \tag{12.26}$$

integers $S_{i,j} \geq 0,\ i, j \in \{1, 2\}$ satisfy (12.23) and that

$$(x_i, u_i) \in X(A, B, S_{i,1}\tau, S_{i,2}\tau),\ i = 1, 2$$

satisfy (12.24). In view of (4.2), (4.3), (12.23) and (12.24), for $i = 1, 2$,

$$I^g(S_{i,1}\tau, S_{i,2}\tau, x_i, u_i) \leq M + a_1 + a\tau S_2. \tag{12.27}$$

Property (P19), (12.23), (12.26) and (12.27) imply that for $i = 1, 2$,

$$|I^f(S_{i,1}\tau, S_{i,2}\tau, x_i, u_i) - I^g(S_{i,1}\tau, S_{i,2}\tau, x_i, u_i)| \leq \varepsilon.$$

By (12.23), (12.27) and property (P20),

$$|x_i(t)| \leq \Delta_0,\ i = 1, 2. \tag{12.28}$$

It follows from (12.25), (12.26) and (12.28) that

$$|h(x_1(S_{1,1}\tau), x_2(S_{2,2}\tau)) - \xi(x_1(S_{1,1}\tau), x_2(S_{2,2}\tau))| \leq \varepsilon.$$

Lemma 4 is proved.

Lemma 5 *Let $h \in \mathfrak{A}_{2n}$, $\widehat{S}_2 > \widehat{S}_1 \geq 1$ be integers, $\varepsilon \in (0, 1)$ and let $(x_*, u_*) \in X(A, B, 0, \infty)$ be an (f, A, B)-overtaking optimal pair and*

$$(\bar{x}_*, \bar{u}_*) \in X(-A, -B, 0, \infty)$$

be an $(\bar{f}, -A, -B)$-overtaking optimal pair. Then there exist a neighborhood $\mathscr{U}$ of f in $\mathscr{M}$ and a neighborhood $\mathscr{V}$ of h in $\mathfrak{A}_{2n}$ such that for each pair of integers $S_{i,2} > S_{i,1} \geq 0$, $i = 1, 2$ satisfying

$$S_{i,2} - S_{i,1} \in [\widehat{S}_1, \widehat{S}_2],\ i = 1, 2, \tag{12.29}$$

each $g \in \mathscr{U}$, each $\xi \in \mathscr{V}$,

$$(y_i, v_i) \in X(A, B, S_{i,1}\tau, S_{i,2}\tau),\ i = 1, 2 \tag{12.30}$$

such that

$$y_1(t) = x_*(t - S_{1,1}\tau),\ v_1(t) = u_*(t - S_{1,1}\tau),\ t \in [S_{1,1}\tau, S_{1,2}\tau], \tag{12.31}$$

$$y_2(t) = \bar{x}_*(S_{2,2}\tau - t),\ v_2(t) = \bar{u}_*(S_{2,2}\tau - t),\ t \in [S_{2,1}\tau, S_{2,2}\tau], \tag{12.32}$$

and each $(x_i, u_i) \in X(A, B, S_{i,1}\tau, S_{i,2}\tau)$, $i = 1, 2$ *satisfying*

$$I^g(S_{1,1}\tau, S_{1,2}\tau, x_1, u_1) + I^g(S_{2,1}\tau, S_{2,2}\tau, x_2, u_2) + \xi(x_1(S_{1,1}\tau), x_2(S_{2,2}\tau))$$

$$\leq I^g(S_{1,1}\tau, S_{1,2}\tau, y_1, v_1) + I^g(S_{2,1}\tau, S_{2,2}\tau, y_2, v_2) + \xi(y_1(S_{1,1}\tau), y_2(S_{2,2}\tau)) + \varepsilon \quad (12.33)$$

the following inequality holds:

$$I^f(S_{1,1}\tau, S_{1,2}\tau, x_1, u_1) + I^f(S_{2,1}\tau, S_{2,2}\tau, x_2, u_2) + h(x_1(S_{1,1}\tau), x_2(S_{2,2}\tau))$$

$$\leq I^f(S_{1,1}\tau, S_{1,2}\tau, y_1, v_1) + I^f(S_{2,1}\tau, S_{2,2}\tau, y_2, v_2) + h(y_1(S_{1,1}\tau), y_2(S_{2,2}\tau)) + 2\varepsilon$$

$$= I^f(0, (S_{1,2} - S_{1,1})\tau, x_*, u_*) + I^{\bar{f}}(0, (S_{2,2} - S_{2,1})\tau, \bar{x}_*, \bar{u}_*) + h(x_*(0), \bar{x}_*(0)) + 2\varepsilon. \quad (12.34)$$

Proof By Theorem 8, there exists $M_0 > 0$ such that

$$|x_*(t)|, |\bar{x}_*(t)| \leq M_0, \ t \in [0, \infty). \quad (12.35)$$

There exists $M_1 > 0$ such that

$$|h(z_1, z_2)| \leq M_1 \text{ for all } (z_1, z_2) \in R^n \times R^n \text{ satisfying } |z_i| \leq M_0, \ i = 1, 2, \quad (12.36)$$

$$|\pi^f(z)|, \ |\pi^{\bar{f}}(z)| \leq M_1 \text{ for all } z \in R^n \text{ satisfying } |z| \leq M_0. \quad (12.37)$$

By Proposition 16, there exists a neighborhood $\mathscr{U}_1$ of f in $\mathscr{M}$ such that the following property holds:

(P21) for each $g \in \mathscr{U}_1$, each pair of integers $S_2 > S_1 \geq 0$ satisfying $S_2 - S_1 \leq \widehat{S}_2$ and each $(x, u) \in X(A, B, S_1\tau, S_2\tau)$ satisfying

$$\min\{I^f(S_1\tau, S_2\tau, x, u), I^g(S_1\tau, S_2\tau, x, u)\} \leq |\mu(f)|\widehat{S}_2\tau + 2M_1$$

we have $|I^f(S_1\tau, S_2\tau, x, u) - I^g(S_1\tau, S_2\tau, x, u)| \leq \varepsilon/16$.

Denote by $\mathscr{V}_1$ the set of all $\xi \in \mathfrak{A}_{2n}$ such that

$$|\xi(z_1, z_2) - h(z_1, z_2)| \leq \varepsilon/16 \text{ for all } z_1, z_2 \in R^n \text{ satisfying } |z_i| \leq M_0, \ i = 1, 2. \quad (12.38)$$

By Lemma 4, there exist a neighborhood $\mathscr{U} \subset \mathscr{U}_1$ of f in $\mathscr{M}$ and a neighborhood $\mathscr{V} \subset \mathscr{V}_1$ of h in $\mathfrak{A}_{2n}$ such that the following property holds:

(P22) for each $g \in \mathscr{U}$, each $\xi \in \mathscr{V}$, each pair of integers $S_{i,2} > S_{i,1} \geq 0$, $i = 1, 2$ satisfying

$$S_{i,2} - S_{i,1} \in [\widehat{S}_1, \widehat{S}_2], \ i = 1, 2,$$

each $(x_1, u_1) \in X(A, B, S_{1,1}\tau, S_{1,2}\tau)$ and each

$$(x_2, u_2) \in X(A, B, S_{2,1}\tau, S_{2,2}\tau)$$

which satisfy

$$I^g(S_{1,1}\tau, S_{1,2}\tau, x_1, u_1) + I^g(S_{2,1}\tau, S_{2,2}\tau, x_2, u_2) + \xi(x_1(S_{1,1}\tau), x_2(S_{2,2}\tau))$$
$$\leq 2|\mu(f)\tau\widehat{S}_2 + 5M_1 + 4$$

we have

$$|I^g(S_{i,1}\tau, S_{i,2}\tau, x_i, u_i) - I^f(S_{i,1}\tau, S_{i,2}\tau, x_i, u_i)| \leq \varepsilon/16, \ i = 1, 2,$$
$$|h(x_1(S_{1,1}\tau), x_2(S_{2,2}\tau)) - \xi(x_1(S_{1,1}\tau), x_2(S_{2,2}\tau))| \leq \varepsilon/16. \tag{12.39}$$

Assume that integers $S_{i,2} > S_{i,1} \geq 0$, $i = 1, 2$ satisfy (12.29),

$$g \in \mathscr{U}, \ \xi \in \mathscr{V}, \tag{12.40}$$

Equations (12.30)–(12.32) hold, $(x_i, u_i) \in X(A, B, S_{i,1}\tau, S_{i,2}\tau)$, $i = 1, 2$ and (12.33) holds. By (3.6), (12.31),(12.32) and Proposition 3,

$$I^f(S_{1,1}\tau, S_{1,2}\tau, y_1, v_1) = I^f(0, (S_{1,2} - S_{1,1})\tau, x_*, u_*)$$
$$= \mu(f)(S_{1,2}\tau - S_{1,1}\tau) + \pi^f(x_*(0)) - \pi^f(x_*(S_{1,2}\tau - S_{1,1}\tau)), \tag{12.41}$$
$$I^f(S_{2,1}\tau, S_{2,2}\tau, y_2, v_2) = I^{\bar{f}}(0, (S_{2,2} - S_{2,1})\tau, \bar{x}_*, \bar{u}_*)$$
$$= \mu(f)(S_{2,2}\tau - S_{2,1}\tau) + \pi^{\bar{f}}(\bar{x}_*(0)) - \pi^{\bar{f}}(\bar{x}_*(S_{2,2}\tau - S_{2,1}\tau)). \tag{12.42}$$

In view of (12.29), (12.35), (12.37), (12.41) and (12.42),

$$I^f(S_{1,1}\tau, S_{1,2}\tau, y_1, v_1), \ I^f(S_{2,1}\tau, S_{2,2}\tau, y_2, v_2) \leq |\mu(f)|\tau\widehat{S}_2 + 2M_1. \tag{12.43}$$

Property (P21), (12.29), (12.40) and (12.43) imply that for $i = 1, 2$,

$$|I^g(S_{i,1}\tau, S_{i,2}\tau, y_i, v_i) - I^f(S_{i,1}\tau, S_{i,2}\tau, y_i, v_i)| \leq \varepsilon/16. \tag{12.44}$$

It follows from ((12.31), (12.32), (12.35) and (12.36) that

$$|h(y_1(S_{1,1}\tau), y_2(S_{2,2}\tau))| \leq M_1. \tag{12.45}$$

By (12.31), (12.32), (12.38) and (12.40),

$$|h(y_1(S_{1,1}\mathcal{T}), y_2(S_{2,2}\mathcal{T})) - \xi(y_1(S_{1,1}\mathcal{T}), y_2(S_{2,2}\mathcal{T}))| \leq \varepsilon/16. \tag{12.46}$$

It follows from (12.33) and (12.43)–(12.46) that

$$\begin{aligned}
&I^g(S_{1,1}\mathcal{T}, S_{1,2}\mathcal{T}, x_1, u_1) + I^g(S_{2,1}\mathcal{T}, S_{2,2}\mathcal{T}, x_2, u_2) + \xi(x_1(S_{1,1}\mathcal{T}), x_2(S_{2,2}\mathcal{T})) \\
&\leq I^f(S_{1,1}\mathcal{T}, S_{1,2}\mathcal{T}, y_1, v_1) + I^f(S_{2,1}\mathcal{T}, S_{2,2}\mathcal{T}, y_2, v_2) \\
&\quad + h(y_1(S_{1,1}\mathcal{T}), y_2(S_{2,2}\mathcal{T})) + 3\varepsilon/16 + \varepsilon \\
&\leq 2|\mu(f)|\mathcal{T}\widehat{S}_2 + 5M_1 + 4.
\end{aligned} \tag{12.47}$$

Property (P22), (12.29), (12.40) and (12.47) imply (12.39). In view of (12.39), (12.41), (12.42) and (12.47),

$$\begin{aligned}
&I^f(S_{1,1}\mathcal{T}, S_{1,2}\mathcal{T}, x_1, u_1) + I^f(S_{2,1}\mathcal{T}, S_{2,2}\mathcal{T}, x_2, u_2) + h(x_1(S_{1,1}\mathcal{T}), x_2(S_{2,2}\mathcal{T})) \\
&\leq I^g(S_{1,1}\mathcal{T}, S_{1,2}\mathcal{T}, x_1, u_1) + I^g(S_{2,1}\mathcal{T}, S_{2,2}\mathcal{T}, x_2, u_2) \\
&\quad + \xi(x_1(S_{1,1}\mathcal{T}), x_2(S_{2,2}\mathcal{T})) + 3\varepsilon/16 \\
&\leq I^f(S_{1,1}\mathcal{T}, S_{1,2}\mathcal{T}, y_1, v_1) + I^f(S_{2,1}\mathcal{T}, S_{2,2}\mathcal{T}, y_2, v_2) \\
&\quad + h(y_1(S_{1,1}\mathcal{T}), y_2(S_{2,2}\mathcal{T})) + 3\varepsilon/16 + \varepsilon + 3\varepsilon/16 \\
&= I^f(0, (S_{1,2} - S_{1,1})\mathcal{T}, x_*, u_*) + I^{\bar{f}}(0, (S_{2,2} - S_{2,1})\mathcal{T}, \bar{x}_*, \bar{u}_*) \\
&\quad + h(x_*(0), \bar{x}_*(0)) + \varepsilon + 3\varepsilon/8.
\end{aligned}$$

Lemma 5 is proved.

13 Proof of Theorem 21

By Lemma 3, there exists $\delta_0 \in (0, \varepsilon)$ such that the following property holds:
(P23) for each $(x_1, u_1) \in X(A, B, 0, L_0\mathcal{T})$ and each

$$(x_2, u_2) \in X(-A, -B, 0, L_0\mathcal{T})$$

which satisfy

$$\psi_h(x_1(0), x_2(0)) \leq \inf(\psi_h) + 4\delta_0,$$

$$I^f(0, L_0\tau, x_1, u_1) - L_0\tau\mu(f) - \pi^f(x_1(0)) + \pi^f(x_1(L_0\tau)) \le 4\delta_0,$$

$$I^{\bar{f}}(0, L_0\tau, x_2, u_2) - L_0\tau\mu(f) - \pi^{\bar{f}}(x_2(0)) + \pi^{\bar{f}}(x_2(L_0\tau)) \le 4\delta_0$$

there exist an (f, A, B)-overtaking optimal pair $(x_1^*, u_1^*) \in X(A, B, 0, \infty)$ and an $(\bar{f}, -A, -B)$-overtaking optimal pair $(x_2^*, u_2^*) \in X(-A, -B, 0, \infty)$ such that

$$\psi_h(x_1^*(0), x_2^*(0)) = \inf(\psi_h)$$

and that for all $t \in [0, L_0\tau]$, $i = 1, 2$,

$$|x_i(t) - x_i^*(t)| \le \varepsilon.$$

In view of Proposition 4, (5.3), (5.4) and the continuity of the functions π^f, $\pi^{\bar{f}}$, there exists $\delta_1 \in (0, \delta_0/4)$ such that: for each $z \in R^n$ satisfying $|z - x_f(0)| \le 2\delta_1$,

$$|\pi^f(z)| = |\pi^f(z) - \pi^f(x_f(0))| \le \delta_0/8,$$

$$|\pi^{\bar{f}}(z)| = |\pi^{\bar{f}}(z) - \pi^{\bar{f}}(x_f(0))| \le \delta_0/8; \tag{13.1}$$

for each $y, z \in R^n$ satisfying

$$|y - x_f(0)| \le 2\delta_1, \ |z - x_f(0)| \le 2\delta_1$$

we have

$$|v(y, z) - \tau\mu(f)| \le \delta_0/8. \tag{13.2}$$

By Theorem 17, there exist an integer $l_0 \ge 1$, $\delta_2 \in (0, \delta_1/8)$, a neighborhood $\mathscr{U}_1$ of f in $\mathscr{M}$ and a neighborhood $\mathscr{V}_1$ of h in $\mathfrak{A}_{2n}$ such that the following property holds:
(P24) for each integer $T > 2l_0$, each $g \in \mathscr{U}_1$, each $\xi \in \mathscr{V}_1$ and each

$$(x, u) \in X(A, B, 0, T\tau)$$

which satisfies $I^g(0, T\tau, x, u) + \xi(x(0), x(T\tau)) \le \sigma(g, \xi, 0, T\tau) + \delta_2$ we have

$$|x(i\tau) - x_f(0)| \le \delta_1 \text{ for all } i = l_0, \dots, T - l_0.$$

By Theorem 2 and Proposition 13, there exist an (f, A, B)-overtaking optimal pair $(x_*, u_*) \in X(A, B, 0, \infty)$ and an $(\bar{f}, -A, -B)$-overtaking optimal pair $(\bar{x}_*, \bar{u}_*) \in X(-A, -B, 0, \infty)$ such that

$$\psi_h(x_*(0), \bar{x}_*(0)) = \inf(\psi_h). \tag{13.3}$$

In view of Theorem 8, the pair (x_*, u_*) is (f, A, B)-good, the pair $(\bar{x}_*, \bar{u}_*)$ is $(\bar{f}, -A, -B)$-good and there exists an integer $l_1 > 0$ such that for all integers $i \geq l_1$,

$$|x_*(i\tau) - x_f(0)| \leq \delta_1, \ |\bar{x}_*(i\tau) - x_f(0)| \leq \delta_1. \tag{13.4}$$

By Proposition 16, there exists a neighborhood $\mathcal{U}_2 \subset \mathcal{U}_1$ of f in $\mathcal{M}$ such that the following property holds:

(P25) for each $g \in \mathcal{U}_2$ and each trajectory-control pair

$$(x, u) \in X(A, B, 0, \tau)$$

which satisfies $\min\{I^f(0, \tau, x, u), I^g(0, \tau, x, u)\} \leq 2 + \tau|\mu(f)|$ the inequality

$$|I^f(0, \tau, x, u) - I^g(0, \tau, x, u)| \leq \delta_1/8$$

holds.

By Lemma 5, there exist a neighborhood $\mathcal{U} \subset \mathcal{U}_2$ of f in $\mathcal{M}$ and a neighborhood $\mathcal{V} \subset \mathcal{V}_1$ of h in $\mathfrak{A}_{2n}$ such that the following property holds:

(P26) for each pair of integers $S_{i,2} > S_{i,1} \geq 0$, $i = 1, 2$ satisfying

$$S_{i,2} - S_{i,1} \leq 2L_0 + 2l_0 + 2l_1 + 4, \ i = 1, 2,$$

each $g \in \mathcal{U}$, each $\xi \in \mathcal{V}$, $(y_i, v_i) \in X(A, B, S_{i,1}\tau, S_{i,2}\tau)$, $i = 1, 2$ such that

$$y_1(t) = x_*(t - S_{1,1}\tau), \ v_1(t) = u_*(t - S_{1,1}\tau), \ t \in [S_{1,1}\tau, S_{1,2}\tau],$$

$$y_2(t) = \bar{x}_*(S_{2,2}\tau - t), \ v_2(t) = \bar{u}_*(S_{2,2}\tau - t), \ t \in [S_{2,1}\tau, S_{2,2}\tau],$$

and each $(x_i, u_i) \in X(A, B, S_{i,1}\tau, S_{i,2}\tau)$, $i = 1, 2$ satisfying

$$I^g(S_{1,1}\tau, S_{1,2}\tau, x_1, u_1) + I^g(S_{2,1}\tau, S_{2,2}\tau, x_2, u_2) + \xi(x_1(S_{1,1}\tau), x_2(S_{2,2}\tau))$$

$$\leq I^g(S_{1,1}\tau, S_{1,2}\tau, y_1, v_1) + I^g(S_{2,1}\tau, S_{2,2}\tau, y_2, v_2) + \xi(y_1(S_{1,1}\tau), y_2(S_{2,2}\tau)) + \delta_0$$

we have

$$-2\delta_0 + I^f(S_{1,1}\tau, S_{1,2}\tau, x_1, u_1)$$

$$+I^f(S_{2,1}\tau, S_{2,2}\tau, x_2, u_2) + h(x_1(S_{1,1}\tau), x_2(S_{2,2}\tau))$$

$$\leq I^f(0, (S_{1,2} - S_{1,1})\tau, x_*, u_*) + I^{\bar{f}}(0, (S_{2,2} - S_{2,1})\tau, \bar{x}_*, \bar{u}_*) + h(x_*(0), \bar{x}_*(0)).$$

Choose $\delta > 0$ and an integer $L_1 > 0$ such that

$$\delta \le 4^{-1}\delta_2(L_0 + l_0 + l_1 + 8)^{-1}, \tag{13.5}$$

$$L_1 > 4L_0 + 4l_0 + 4l_1 + 8. \tag{13.6}$$

Assume that an integer

$$T \ge L_1,\ g \in \mathscr{U},\ \xi \in \mathscr{V},\ (x, u) \in X(A, B, 0, T\tau), \tag{13.7}$$

$$I^g(0, T\tau, x, u) + \xi(x(0), x(T\tau)) \le \sigma(g, \xi, 0, T\tau) + \delta. \tag{13.8}$$

It follows from property (P24) and (13.5)–(13.8) that

$$|x(i\tau) - x_f(0)| \le \delta_1 \text{ for all } i = l_0, \ldots, T - l_0. \tag{13.9}$$

By Proposition 18, (5.1), (13.6) and (13.7), there exists a trajectory-control pair $(\tilde{x}, \tilde{u}) \in X(A, B, 0, T\tau)$ such that

$$\tilde{x}(t) = x_*(t),\ \tilde{u}(t) = u_*(t),\ t \in [0, \tau(L_0 + l_0 + l_1 + 3)],$$

$$\tilde{x}(t) = x(t),\ \tilde{u}(t) = u(t),\ t \in [\tau(L_0 + l_0 + l_1 + 4), \tau(T - l_0 - l_1 - L_0 - 4)],$$

$$\tilde{x}(t) = \bar{x}_*(T\tau - t),\ \tilde{u}(t) = \bar{u}_*(T\tau - t),\ t \in [\tau(T - l_0 - l_1 - L_0 - 3), T\tau],$$

$$I^f(\tau(l_0 + l_1 + L_0 + 3), \tau(l_0 + l_1 + L_0 + 4), \tilde{x}, \tilde{u})$$

$$= v(x_*(\tau(l_0 + l_1 + L_0 + 3)), x(\tau(l_0 + l_1 + L_0 + 4))),$$

$$I^f(\tau(T - l_0 - l_1 - L_0 - 4), \tau(T - l_0 - l_1 - L_0 - 3), \tilde{x}, \tilde{u})$$

$$= v(x(\tau(T - l_0 - l_1 - L_0 - 4)), \bar{x}_*(\tau(l_0 + l_1 + L_0 + 3))). \tag{13.10}$$

By (13.4), (13.6), (13.7), (13.9) and (13.10),

$$|\tilde{x}(\tau(L_0 + l_0 + l_1 + 3)) - x_f(0)| = |x_*(\tau(L_0 + l_0 + l_1 + 3)) - x_f(0)| \le \delta_1,$$

$$|\tilde{x}(\tau(L_0 + l_0 + l_1 + 4)) - x_f(0)| = |x(\tau(L_0 + l_0 + l_1 + 4)) - x_f(0)| \le \delta_1,$$

$$|\tilde{x}(\tau(T - L_0 - l_0 - l_1 - 4)) - x_f(0)| = |x(\tau(T - L_0 - l_0 - l_1 - 4)) - x_f(0)| \le \delta_1,$$

$$|\tilde{x}(\tau(T - L_0 - l_0 - l_1 - 3)) - x_f(0)| = |\bar{x}_*(\tau(L_0 + l_0 + l_1 + 3)) - x_f(0)| \le \delta_1. \tag{13.11}$$

It follows from (13.8) and (13.10) that

$$-\delta \leq I^g(0, T\tau, \tilde{x}, \tilde{u}) + \xi(\tilde{x}(0), \tilde{x}(T\tau))$$

$$-(I^g(0, T\tau, x, u) - \xi(x(0), x(T\tau)))$$

$$= I^g(0, \tau(L_0 + l_0 + l_1 + 3), \tilde{x}, \tilde{u}) + I^g(\tau(L_0 + l_0 + l_1 + 3), \tau(L_0 + l_0 + l_1 + 4), \tilde{x}, \tilde{u})$$

$$+ I^g(\tau(T - L_0 - l_0 - l_1 - 4), \tau(T - L_0 - l_0 - l_1 - 3), \tilde{x}, \tilde{u})$$

$$+ I^g(\tau(T - l_0 - l_1 - L_0 - 3), T\tau, \tilde{x}, \tilde{u}) + \xi(\tilde{x}(0), \tilde{x}(\tau T))$$

$$- I^g(0, \tau(L_0 + l_0 + l_1 + 3), x, u) - I^g(\tau(L_0 + l_0 + l_1 + 3), \tau(L_0 + l_0 + l_1 + 4), x, u)$$

$$- I^g(\tau(T - L_0 - l_0 - l_1 - 4), \tau(T - L_0 - l_0 - l_1 - 3), x, u)$$

$$- I^g(\tau(T - l_0 - l_1 - L_0 - 3), \tau T, x, u) - \xi(x(0), x(\tau T)). \tag{13.12}$$

We will estimate

$$I^g(\tau(L_0 + l_0 + l_1 + 3), \tau(L_0 + l_0 + l_1 + 4), \tilde{x}, \tilde{u})$$

$$- I^g(\tau(L_0 + l_0 + l_1 + 3), \tau(L_0 + l_0 + l_1 + 4), x, u),$$

$$I^g(\tau(T - L_0 - l_0 - l_1 - 4), \tau(T - L_0 - l_0 - l_1 - 3), \tilde{x}, \tilde{u})$$

$$- I^g(\tau(T - L_0 - l_0 - l_1 - 4), \tau(T - L_0 - l_0 - l_1 - 3), x, u).$$

It follows from (13.10), (13.11) and the choice of δ_1 (see (13.2)) that

$$I^f(\tau(L_0 + l_0 + l_1 + 3), \tau(L_0 + l_0 + l_1 + 4), \tilde{x}, \tilde{u})$$

$$= v(x_*(\tau(l_0 + l_1 + L_0 + 3)), x(\tau(l_0 + l_1 + L_0 + 4))) \leq \tau\mu(f) + \delta_0/8,$$

$$I^f(\tau(T - L_0 - l_0 - l_1 - 4), \tau(T - L_0 - l_0 - l_1 - 3), \tilde{x}, \tilde{u})$$

$$= v(x(\tau(T - l_0 - l_1 - L_0 - 4)), \bar{x}_*(\tau(l_0 + l_1 + L_0 + 3))) \leq \tau\mu(f) + \delta_0/8.$$

Combined with property (P25) and (13.7) these inequalities imply that

$$I^g(\tau(L_0 + l_0 + l_1 + 3), \tau(L_0 + l_0 + l_1 + 4), \tilde{x}, \tilde{u})$$

$$\leq I^f(\tau(L_0 + l_0 + l_1 + 3), \tau(L_0 + l_0 + l_1 + 4), \tilde{x}, \tilde{u}) + \delta_1/8 \leq \tau\mu(f) + \delta_0/8 + \delta_1/8, \tag{13.13}$$

$$I^g(\tau(T - l_0 - l_1 - L_0 - 4), \tau(T - l_0 - l_1 - L_0 - 3), \tilde{x}, \tilde{u})$$

$$\leq I^f(\tau(T - l_0 - l_1 - L_0 - 4), \tau(T - l_0 - l_1 - L_0 - 3), \tilde{x}, \tilde{u}) + \delta_1/8$$

$$\leq \tau\mu(f) + \delta_0/8 + \delta_1/8. \quad (13.14)$$

By (13.6), (13.7), (13.9) and the choice of δ_1 (see (13.2)),

$$I^f(\tau(L_0 + l_0 + l_1 + 3), \tau(L_0 + l_0 + l_1 + 4), x, u)$$

$$\geq v(x(\tau(L_0 + l_0 + l_1 + 3)), x(\tau(L_0 + l_0 + l_1 + 4))) \geq \tau\mu(f) - \delta_0/8, \quad (13.15)$$

$$I^f(\tau(T - L_0 - l_0 - l_1 - 4), \tau(T - L_0 - l_0 - l_1 - 3), x, u)$$

$$\geq v(x(\tau(T - L_0 - l_0 - l_1 - 4)), x(\tau(T - L_0 - l_0 - l_1 - 3))) \geq \tau\mu(f) - \delta_0/8. \quad (13.16)$$

We show that

$$I^g(\tau(L_0 + l_0 + l_1 + 3), \tau(L_0 + l_0 + l_1 + 4), x, u) \geq \tau\mu(f) - \delta_0/4. \quad (13.17)$$

Assume the contrary. Then

$$I^g(\tau(L_0 + l_0 + l_1 + 3), \tau(L_0 + l_0 + l_1 + 4), x, u) < \tau\mu(f) - \delta_0/4.$$

Together with property (P25) and (13.7) this implies that

$$I^f(\tau(L_0 + l_0 + l_1 + 3), \tau(L_0 + l_0 + l_1 + 4), x, u)$$

$$\leq I^g(\tau(L_0 + l_0 + l_1 + 3), \tau(L_0 + l_0 + l_1 + 4), x, u) + \delta_1/8$$

$$< \tau\mu(f) - \delta_0/4 + \delta_1/8 < \mu(f) - \delta_0/8.$$

This contradicts (13.15). The contradiction we have reached proves (13.17).

We show that

$$I^g(\tau(T - L_0 - l_0 - l_1 - 4), \tau(T - L_0 - l_0 - l_1 - 3), x, u) \geq \tau\mu(f) - \delta_0/4. \quad (13.18)$$

Assume the contrary. Then

$$I^g(\tau(T - L_0 - l_0 - l_1 - 4), \tau(T - L_0 - l_0 - l_1 - 3), x, u) < \tau\mu(f) - \delta_0/4.$$

Together with property (P25) and (13.7) this implies that

$$I^f(\tau(T - L_0 - l_0 - l_1 - 4), \tau(T - L_0 - l_0 - l_1 - 3), x, u)$$

$$\leq I^g(\tau(T - L_0 - l_0 - l_1 - 4), \tau(T - L_0 - l_0 - l_1 - 3), x, u) + \delta_1/8$$

$$< \tau\mu(f) - \delta_0/4 + \delta_1/8 < \tau\mu(f) - \delta_0/8.$$

This contradicts (13.16). The contradiction we have reached proves (13.18). In view of (13.13), (13.14), (13.17) and (13.18),

$$I^g(\tau(L_0 + l_0 + l_1 + 3), \tau(L_0 + l_0 + l_1 + 4), \tilde{x}, \tilde{u})$$

$$- I^g(\tau(L_0 + l_0 + l_1 + 3), \tau(L_0 + l_0 + l_1 + 4), x, u) \leq \delta_0/8 + \delta_1/8 + \delta_0/4, \tag{13.19}$$

$$I^g(\tau(T - L_0 - l_0 - l_1 - 4), \tau(T - L_0 - l_0 - l_1 - 3), \tilde{x}, \tilde{u})$$

$$- I^g(\tau(T - L_0 - l_0 - l_1 - 4), \tau(T - L_0 - l_0 - l_1 - 3), x, u) \leq \delta_0/8 + \delta_1/8 + \delta_0/4. \tag{13.20}$$

By (13.12), (13.19) and (13.20),

$$I^g(0, \tau(L_0 + l_0 + l_1 + 3), x, u)$$

$$+I^g(\tau(T - l_0 - l_1 - L_0 - 3)), \tau T, x, u) - \xi(x(0), x(\tau T))$$

$$\leq \delta + I^g(0, \tau(L_0 + l_0 + l_1 + 3), \tilde{x}, \tilde{u}) + I^g(\tau(T - l_0 - l_1 - L_0 - 3), \tau T, \tilde{x}, \tilde{u})$$

$$+ \xi(\tilde{x}(0), \tilde{x}(\tau T)) + \delta_1/4 + 3\delta_0/4. \tag{13.21}$$

Property (P26), Proposition 3, (13.7), (13.10) and (13.21) imply that

$$I^f(0, \tau(L_0 + l_0 + l_1 + 3), x, u) + I^f(\tau(T - l_0 - l_1 - L_0 - 3), \tau T, x, u)$$

$$+h(x(0), x(\tau T))$$

$$\leq 2\delta_0 + I^f(0, \tau(L_0 + l_0 + l_1 + 3), x_*, u_*)$$

$$+I^{\bar{f}}(0, \tau(L_0 + l_0 + l_1 + 3), \bar{x}_*, \bar{u}_*) + h(x_*(0), \bar{x}_*(0))$$

$$\leq 2\delta_0 + 2\tau(L_0 + l_0 + l_1 + 3)\mu(f) + h(x_*(0), \bar{x}_*(0))$$

$$+\pi^f(x_*(0)) - \pi^f(x_*(\tau(L_0 + l_0 + l_1 + 3))) + \pi^{\bar{f}}(\bar{x}_*(0))$$

$$- \pi^{\bar{f}}(\bar{x}_*(\tau(L_0 + l_0 + l_1 + 3))). \tag{13.22}$$

In view of (13.1), (13.4), (13.6), (13.8) and (13.9),

$$|\pi^f(x_*(\tau(L_0+l_0+l_1+3)))|,\ |\pi^{\bar f}(\bar x_*(\tau(L_0+l_0+l_1+3)))| \le \delta_0/8,$$

$$|\pi^f(x(\tau(L_0+l_0+l_1+3)))|,\ |\pi^{\bar f}(x(T\tau-\tau(L_0+l_0+l_1+3)))| \le \delta_0/8. \tag{13.23}$$

By (13.22) and (13.23),

$$I^f(0,\tau(L_0+l_0+l_1+3),x,u)$$
$$+I^f(\tau(T-l_0-l_1-L_0-3),\tau T,x,u)+h(x(0),x(\tau T))$$
$$\le 2\tau(L_0+l_0+l_1+3)\mu(f)+\psi_h(x_*(0),\bar x_*(0))+2\delta_0+\delta_0/4. \tag{13.24}$$

Set

$$\widehat{x}(t)=x(T\tau-t),\ \widehat{u}(t)=u(T\tau-t),\ t\in[0,T\tau]. \tag{13.25}$$

In view of (3.6) and (13.25),

$$I^f(\tau(T-l_0-l_1-L_0-3),T\tau,x,u)=I^{\bar f}(0,\tau(L_0+l_0+l_1+3),\widehat{x},\widehat{u}). \tag{13.26}$$

By (12.1) and (13.23)–(13.26),

$$2\delta_0+\delta_0/4+\psi_h(x_*(0),\bar x_*(0))$$
$$\ge I^f(0,\tau(L_0+l_0+l_1+3),x,u)-\tau\mu(f)(L_0+l_0+l_1+3)$$
$$+I^{\bar f}(0,\tau(L_0+l_0+l_1+3),\widehat{x},\widehat{u})-\tau\mu(f)(L_0+l_0+l_1+3)+h(x(0),\widehat{x}(0))$$
$$=\psi_h(x(0),\widehat{x}(0))+I^f(0,\tau(L_0+l_0+l_1+3),x,u)-\pi^f(x(0))$$
$$-\tau\mu(f)(L_0+l_0+l_1+3)$$
$$+\pi^f(x(\tau(L_0+l_0+l_1+3)))-\pi^f(x(\tau(L_0+l_0+l_1+3)))$$
$$+I^{\bar f}(0,\tau(L_0+l_0+l_1+3),\widehat{x},\widehat{u})-\tau\mu(f)(L_0+l_0+l_1+3)$$
$$-\pi^{\bar f}(\widehat{x}(0))+\pi^{\bar f}(\widehat{x}(\tau(L_0+l_0+l_1+3)))-\pi^{\bar f}(\widehat{x}(\tau(L_0+l_0+l_1+3)))$$
$$\ge\psi_h(x(0),\widehat{x}(0))+I^f(0,\tau(L_0+l_0+l_1+3),x,u)$$
$$-\tau\mu(f)(L_0+l_0+l_1+3)-\pi^f(x(0))+\pi^f(x(\tau(L_0+l_0+l_1+3)))$$

$$+I^{\bar{f}}(0, \tau(L_0+l_0+l_1+3), \widehat{x}, \widehat{u}) - \tau\mu(f)(L_0+l_0+l_1+3)$$

$$-\pi^{\bar{f}}(\widehat{x}(0)) + \pi^{\bar{f}}(\widehat{x}(\tau(L_0+l_0+l_1+3))) - \delta_0/4.$$

Together with (13.3) this implies that

$$\psi_h(x(0), \widehat{x}(0)) + I^f(0, \tau(L_0+l_0+l_1+3), x, u)$$

$$-\tau\mu(f)(L_0+l_0+l_1+3) - \pi^f(x(0)) + \pi^f(x(\tau(L_0+l_0+l_1+3)))$$

$$+I^{\bar{f}}(0, \tau(L_0+l_0+l_1+3), \widehat{x}, \widehat{u}) - \tau\mu(f)(L_0+l_0+l_1+3)$$

$$-\pi^{\bar{f}}(\widehat{x}(0)) + \pi^{\bar{f}}(\widehat{x}(\tau(L_0+l_0+l_1+3))) \le 3\delta_0 + \inf(\psi_h).$$

Combined with Proposition 2 this implies that

$$\psi_h(x(0), \widehat{x}(0)) \le \inf(\psi_h) + 3\delta_0, \tag{13.27}$$

$$I^f(0, \tau L_0, x, u) - \tau\mu(f)L_0 - \pi^f(x(0)) + \pi^f(x(L_0\tau))$$

$$\le I^f(0, \tau(L_0+l_0+l_1+3), x, u) - \tau\mu(f)(L_0+l_0+l_1+3)$$

$$-\pi^f(x(0)) + \pi^f(x(\tau(L_0+l_0+l_1+3))) \le 3\delta_0, \tag{13.28}$$

$$I^{\bar{f}}(0, \tau L_0, \widehat{x}, \widehat{u}) - \tau\mu(f)L_0 - \pi^{\bar{f}}(\widehat{x}(0)) + \pi^{\bar{f}}(\widehat{x}(\tau L_0))$$

$$\le I^{\bar{f}}(0, \tau(L_0+l_0+l_1+3), \widehat{x}, \widehat{u}) - \tau\mu(f)(L_0+l_0+l_1+3)$$

$$-\pi^{\bar{f}}(\widehat{x}(0)) + \pi^{\bar{f}}(\widehat{x}(\tau(L_0+l_0+l_1+3))) \le 3\delta_0. \tag{13.29}$$

By (13.25), (13.27)–(13.29) and property (P23), there are an (f, A, B)-overtaking optimal pair $(\xi_1, \eta_1) \in X(A, B, 0, \infty)$ and an $(\bar{f}, -A, -B)$-overtaking optimal pair $(\xi_2, \eta_2) \in X(-A, -B, 0, \infty)$ such that $\psi_h(\xi_1(0), \xi_2(0)) = \inf(\psi_h)$ and that for all $t \in [0, L_0\tau]$,

$$\varepsilon \ge |x(t) - \xi_1(t)|, \; \varepsilon \ge |\xi_2(t) - \widehat{x}(t)| = |\xi_2(t) - x(T\tau - t)|.$$

Theorem 21 is proved.

14 Genericity Results

We use the notation, definitions and assumptions introduced in Sects. 1–4. For each nonempty set X and each function $h : X \to R^1$ set $\inf(h) = \inf\{h(x) : x \in X\}$. We prove the following results.

Theorem 22 *There exists an everywhere dense set $\mathscr{B} \subset \mathfrak{A}_n$ which is a countable intersection of open subsets of $\mathfrak{A}_n$ such that for every $h \in \mathscr{B}$ the following assertions hold:*

(1) $\inf(\pi^{\bar f} + h)$ is finite and attained at a unique point $\bar x \in R^n$.
(2) for every $\varepsilon > 0$ there are a neighborhood $\mathscr{V}$ of h in $\mathfrak{A}_n$ and $\delta > 0$ such that for each $\xi \in \mathscr{V}$, $\inf(\xi + \pi^{\bar f})$ is finite and if $z \in R^n$ satisfies $(\xi + \pi^{\bar f})(z) \le \inf(\xi + \pi^{\bar f}) + \delta$, then

$$|z - \bar x| \le \varepsilon,\ |(\xi + \pi^{\bar f})(z) - (h + \pi^{\bar f})(\bar x)| \le \varepsilon.$$

Theorem 23 *There exists an everywhere dense set $\mathscr{B} \subset \mathfrak{A}_{2n}$ which is a countable intersection of open subsets of $\mathfrak{A}_{2n}$ such that for every $h \in \mathscr{B}$ the following assertions hold:*

(1) $\inf(\psi_h)$ is finite and attained at a unique point $\bar x = (\bar x_1, \bar x_2) \in R^n \times R^n$;
(2) for every $\varepsilon > 0$ there are a neighborhood $\mathscr{V}$ of h in $\mathfrak{A}_{2n}$ and $\delta > 0$ such that for each $\xi \in \mathscr{V}$, $\inf(\psi_\xi)$ is finite and if $z = (z_1, z_2) \in R^n \times R^n$ satisfies $(\psi_\xi)(z) \le \inf(\psi_\xi) + \delta$, then

$$|z_i - \bar x_i| \le \varepsilon,\ i = 1, 2,\ |\psi_\xi(z) - \psi_h(\bar x)| \le \varepsilon.$$

Theorems 22 and 23 follow from Propositions 7 and 8 and the following result.

Theorem 24 *Let $k \ge 1$ be an integer and $g : R^k \to R^1$ be a continuous function such that $\lim_{|z|\to\infty} g(z) = \infty$. There exists an everywhere dense set $\mathscr{B} \subset \mathfrak{A}_k$ which is a countable intersection of open subsets of $\mathfrak{A}_k$ such that for every $h \in \mathscr{B}$ the following assertions hold:*

(1) $\inf(g + h)$ is finite and attained at a unique point $\bar x \in R^n$.
(2) for every $\varepsilon > 0$ there are a neighborhood $\mathscr{V}$ of h in $\mathfrak{A}_k$ and $\delta > 0$ such that for each $\xi \in \mathscr{V}$, $\inf(g + \xi)$ is finite and if $z \in R^k$ satisfies $(g + \xi)(z) \le \inf(g + \xi) + \delta$, then

$$|z - \bar x| \le \varepsilon,\ |(g + \xi)(z) - (g + h)(\bar x)| \le \varepsilon.$$

Proof We obtain our result as a realization of the variational principle (see Theorem 4.1 of [23]). Let d be the metric in $\mathfrak{A}_k$ induced by its uniformity. By Theorem 4.1 of [23] it is sufficient to show that the following property holds:

(H) for any $h \in \mathfrak{A}_k$, any $\varepsilon > 0$ and any $\gamma > 0$ there exist a nonempty open set $\mathscr{W}$ in $\mathfrak{A}_k$, $x \in R^k$, $\alpha \in R^1$ and $\eta > 0$ such that

$$\mathscr{W} \cap \{b \in \mathfrak{A}_k : \ d(a, b) < \varepsilon\} \neq \emptyset$$

and for each $\xi \in \mathscr{W}$,

(1) $\inf(g + \xi)$ is finite;
(2) if $z \in R^k$ is such that $(g + \xi)(z) \leq \inf(g + \xi) + \eta$, then $|z - x| \leq \gamma$ and $|(g + \xi)(z) - \alpha| \leq \gamma$.

Let $h \in \mathfrak{A}_k$, $\varepsilon > 0$ and $\gamma > 0$. There exists $\bar{x} \in R^n$ such that

$$(g + h)(\bar{x}) = \inf(g + h). \tag{14.1}$$

For each integer $i \geq 1$ define

$$h_i(z) = h(z) + i^{-1} \min\{1, |z - \bar{x}|\}, \ z \in R^k. \tag{14.2}$$

Clearly, for every integer $i \geq 0$, $h_i \in \mathfrak{A}_k$,

$$\{z \in R^k : \ (g + h_i)(z) = \inf(g + h_i)\} = \{\bar{x}\}, \tag{14.3}$$

$$\lim_{i \to \infty} h_i = h. \tag{14.4}$$

In view of (14.4), there exists a natural number j such that

$$d(h_j, h) < \varepsilon/2. \tag{14.5}$$

Choose a number $M > 0$ such that

$$|\bar{x}| < M, \tag{14.6}$$

$$\{z \in R^k : \ g(z) \leq a_1 + (g + h)(\bar{x}) + 2\} \subset \{z \in R^k : \ |z| \leq M\} \tag{14.7}$$

and $\delta, \eta \in (0, 1)$ such that

$$3\delta j < \min\{\gamma, 1\}, \tag{14.8}$$

$$\eta \in (0, \delta). \tag{14.9}$$

Set

$$\alpha = (g + h)(\bar{x}). \tag{14.10}$$

Denote by $\mathscr{W}$ an open neighborhood of h_j in $\mathfrak{A}_k$ such that

$$\mathscr{W} \subset \{\xi \in \mathfrak{A}_k : \ |\xi(z) - h_j(z)| \leq \delta \text{ for all } z \in R^k \text{ such that } |z| \leq M\}. \tag{14.11}$$

In view of (14.5) and (14.11),

$$h_j \in \mathscr{W} \cap \{\xi \in \mathfrak{A}_k : d(\xi, h) < \varepsilon\} \neq \emptyset. \tag{14.12}$$

Let

$$\xi \in \mathscr{W}. \tag{14.13}$$

Clearly, $\inf(g + \xi)$ is finite. By (14.1), (14.6), (14.11) and (14.13),

$$\inf(g + \xi) \leq (g + \xi)(\bar{x}) \leq (g + h_j)(\bar{x}) + \delta = \inf(g + h) + \delta. \tag{14.14}$$

Assume that $z \in R^k$ satisfies

$$(g + \xi)(z) \leq \inf(g + \xi) + \eta. \tag{14.15}$$

By (14.2), (14.9), (14.14) and (14.15),

$$(g + \xi)(z) \leq (g + h_j)(\bar{x}) + \delta + \eta \leq (g + h)(\bar{x}) + 2\delta. \tag{14.16}$$

In view of (14.2), (4.3) and (14.16),

$$g(z) \leq (g + h)(\bar{x}) + 2 + a_1. \tag{14.17}$$

Relations (14.7) and (14.17) imply that

$$|z| \leq M. \tag{14.18}$$

It follows from (11.11), (14.13) and (14.18) that

$$|\xi(z) - h_j(z)| \leq \delta, \ |(g + \xi)(z) - (g + h_j)(z)| \leq \delta. \tag{14.19}$$

By (14.1), (14.2), (14.19), (14.14) and (14.15),

$$\inf(g + h) + \delta + \eta \geq \inf(g + \xi) + \eta \geq (g + \xi)(z)$$

$$\geq (g + h_j)(z) - \delta \geq (g + h_j)(\bar{x}) - \delta = (g + h)(\bar{x}) - \delta. \tag{14.20}$$

By (14.1), (14.8) and (14.20),

$$|(g + \xi)(z) - (g + h)(\bar{x})| \leq \delta + \eta \leq 2\delta < \gamma. \tag{14.21}$$

It follows from (14.9), (14.14), (14.15) and (14.19) that

$$(g + h_j)(z) \leq (g + \xi)(z) + \delta \leq \inf(g + \xi) + 2\delta$$
$$\leq \inf(g + h) + 3\delta \leq (g + h)(z) + 3\delta.$$

Together with (14.2) and (14.8) this implies that $j^{-1} \min\{1, |z - \bar{x}|\} \leq 3\delta$, $|z - \bar{x}| \leq 3\delta j < \gamma$. Thus (H) holds. This completes the proof of Theorem 24.

References

1. Anderson BDO, Moore JB (1971) Linear optimal control. Prentice-Hall, Englewood Cliffs
2. Aseev SM, Veliov VM (2012) Maximum principle for infinite-horizon optimal control problems with dominating discount. Dyn Contin Discret Impuls Syst SERIES B 19:43–63
3. Aubry S, Le Daeron PY (1983) The discrete Frenkel-Kontorova model and its extensions I. Phys D 8:381–422
4. Blot J, Hayek N (2014) Infinite-horizon optimal control in the discrete-time framework. SpringerBriefs in optimization. Springer, New York
5. Bright I (2012) A reduction of topological infinite-horizon optimization to periodic optimization in a class of compact 2-manifolds. J Math Anal Appl 394:84–101
6. Carlson DA, Haurie A, Leizarowitz A (1991) Infinite horizon optimal control. Springer, Berlin
7. Cartigny P, Michel P (2003) On a sufficient transversality condition for infinite horizon optimal control problems. Autom J IFAC 39:1007–1010
8. Damm T, Grune L, Stieler M, Worthmann K (2014) An exponential turnpike theorem for dissipative discrete time optimal control problems. SIAM J Control Optim 52:1935–1957
9. De Oliveira VA, Silva GN (2009) Optimality conditions for infinite horizon control problems with state constraints. Nonlinear Anal 71:1788–1795
10. Gaitsgory V, Grune L, Thatcher N (2015) Stabilization with discounted optimal control. Syst Control Lett 82:91–98
11. Jasso-Fuentes H, Hernandez-Lerma O (2008) Characterizations of overtaking optimality for controlled diffusion processes. Appl Math Optim 57:349–369
12. Khlopin DV (2013) Necessary conditions of overtaking equilibrium for infinite horizon differential games. Mat Teor Igr Pril 5:105–136
13. Leizarowitz A, Mizel VJ (1989) One dimensional infinite horizon variational problems arising in continuum mechanics. Arch Ration Mech Anal 106:161–194
14. Makarov VL, Rubinov AM (1977) Mathematical theory of economic dynamics and equilibria. Springer, New York
15. Mammadov M (2014) Turnpike theorem for an infinite horizon optimal control problem with time delay. SIAM J Control Optim 52:420–438
16. Marcus M, Zaslavski AJ (1999) The structure of extremals of a class of second order variational problems. Ann Inst H Poincaré Anal non Lineare **16**, 593–629 (1999)
17. McKenzie LW (1976) Turnpike theory. Econometrica 44:841–866
18. Mordukhovich BS (2011) Optimal control and feedback design of state-constrained parabolic systems in uncertainly conditions. Appl Anal 90:1075–1109
19. Pickenhain S, Lykina V, Wagner M (2008) On the lower semicontinuity of functionals involving Lebesgue or improper Riemann integrals in infinite horizon optimal control problems. Control Cybern 37:451–468
20. Samuelson PA (1965) A catenary turnpike theorem involving consumption and the golden rule. Am Econ Rev 55:486–496
21. Trelat E, Zuazua E (2015) The turnpike property in finite-dimensional nonlinear optimal control. J Differ Equ 218:81–114
22. Zaslavski AJ (2006) Turnpike properties in the calculus of variations and optimal control. Springer, New York
23. Zaslavski AJ (2010) Optimization on metric and normed spaces. Springer optimization and its applications. Springer, New York
24. Zaslavski AJ (2011) The existence and structure of approximate solutions of dynamic discrete time zero-sum games. J Nonlinear Convex Anal 12:49–68

25. Zaslavski AJ (2014) Turnpike phenomenon and infinite horizon optimal control. Springer optimization and its applications. Springer, New York
26. Zaslavski AJ (2015) Turnpike theory of continuous-time linear optimal control problems. Springer optimization and its applications. Springer, Cham (2015)
27. Zaslavski AJ (2016) Structure of solutions of optimal control problems on large intervals: a survey of recent results. Pure Appl Funct Anal 1:123–158
28. Zaslavski AJ, Leizarowitz A (1998) Optimal solutions of linear periodic control systems with convex integrands. Appl Math Optim 37:127–150

Index

A
Approximate optimal trajectories, 101
Atomless Loeb probability space, 56
Atomless probability, 59
Atomless probability structure, 64

B
Bayesian game, 47, 51, 54, 55, 65, 69
Biting Lemma, 2
Bolza optimal control problems, 114
Boundary conditions, 2
Bounded variation, 3
Brownian motion, 94

C
Carathéodory, 60
Carathéodory's procedure, 61
Carathéodory's theorem, 53
Continuous adapted process, 94
Convergence of approximate solutions, 117
Countably subadditive, 60, 61

D
D-correspondence, 53
Decomposing set, 60
Dialectic, 52
Diffused private information structure, 59
Disparate, 50
Dispersed, 50
Disturbed (perturbed) games, 49
D-property, 48, 51, 53, 54, 58, 59, 64, 70
  with respect to a measurable, 53
  with respect to a measure-preserving map, 52

E
Epiconvergence, 2
Equivalence theorem of Aumann, 49
Extended Lebesgue interval, 47

G
Good trajectory-control pairs, 102
Gronwall's inequality, 96

H
Hausdorff metric, 78
HJ function, 76

I
Independent type, 55
Induced distribution, 64
Infinite recursion, 52

K
Kakutani, 53
KRS counterexample, 53
KRS example, 52

S. Kusuoka and T. Maruyama (eds.), *Advances in Mathematical Economics*, Advances in Mathematical Economics 21,
DOI 10.1007/978-981-10-4145-7

KRS-like, 59
KRS-like game, 51–54, 57, 58, 66, 67, 69

L
Lebesgue extension, 53, 54, 59, 63–71
Lebesgue interval, 53
Lebesgue measure, 57, 59
Linear control systems, 101

M
Martingales, 89
Measure-preserving function, 53
Measure preserving map, 59, 64
Mixed-strategy Nash equilibrium, 48
Monotonic, 60
m-th fold Lebesgue extension, 64

N
n-fold Lebesgue extension, 64
Non-atomic measure spaces, 49
Non-cooperative game theory, 49
Nowhere equivalence, 48, 67
Nowhere equivalence σ-algebras, 67
Nowhere equivalent, 69

O
Outer measure, 60, 62
Overtaking optimal trajectory-control pair, 102

P
Partition, 62
Polish space, 75
Private information game, 55
Private information structure, 56
Progressively measurable, 76
Progressively measurable processs, 89
Prohorov metric, 75
Pure strategy of a Bayesian game, 55
Pure-strategy equilibrium, 59
Pure-strategy Nash equilibrium (PSNE), 47, 48, 52–56, 58, 59, 64–70
Purification, 49, 50

R
Recursive upgrading, 52
Relative diffuseness, 54
Bayesian games, 56
Relative d-property, 59, 64
Relatively diffused, 56
Relative private information structure, 56

S
Saturated, 51
Saturated probability space, 51, 56, 69, 70
Saturated space, 54, 64
Saturation property, 47, 48
Second order evolution inclusion, 2
Semi-martingales, 89
Setwise coarser, 56, 64
Stability of the turnpike phenomenon, 101
Subdifferential operators, 2
Subjective beliefs, 49
Super-atomless, 51
Supermartingale problem, 76
Symmetrization, 49

T
Trajectory-control pair, 100, 150
Turnpike property, 101
Two independent σ-algebras, 62

U
Uncountable cardinality, 62
Uniformity, 106

V
Variational limits, 2
Variational principle, 156

Y
Young measures, 2

Printed in the United States
By Bookmasters